Astronomers' Universe

The Astronomers' Universe series attracts scientifically curious readers with a passion for astronomy and its related fields. In this series, you will venture beyond the basics to gain a deeper understanding of the cosmos—all from the comfort of your chair.

Our books cover any and all topics related to the scientific study of the Universe and our place in it, exploring discoveries and theories in areas ranging from cosmology and astrophysics to planetary science and astrobiology.

This series bridges the gap between very basic popular science books and higher-level textbooks, providing rigorous, yet digestible forays for the intrepid lay reader. It goes beyond a beginner's level, introducing you to more complex concepts that will expand your knowledge of the cosmos. The books are written in a didactic and descriptive style, including basic mathematics where necessary.

More information about this series at http://www.springer.com/series/6960

John C. Barentine

Mystery of the Ashen Light of Venus

Investigating a 400-Year-Old Phenomenon

John C. Barentine
International Dark-Sky Association
Tucson, AZ, USA

ISSN 1614-659X ISSN 2197-6651 (electronic)
Astronomers' Universe
ISBN 978-3-030-72714-7 ISBN 978-3-030-72715-4 (eBook)
https://doi.org/10.1007/978-3-030-72715-4

Cover illustration: A painting by astronomer William K. Hartmann, who, with fellow astronomer Dale P. Cruikshank, followed Venus telescopically for several days as it passed between Earth and Sun. On one those days, they both thought they could see a faint discoloration of Venus's dark side, as depicted in this painting. © W. K. Hartmann, Senior Scientist Emeritus, Planetary Science Institute, Tucson

This Springer imprint is published by the registered company Springer Nature Switzerland AG
The registered company address is: Gewerbestrasse 11, 6330 Cham, Switzerland

This book is dedicated to the memory of Richard Baum (1930–2017), a fine amateur astronomer whose contributions to the history of its subject over many decades have so informed my work.

Acknowledgments

As much as it is a cliché, it is worth reiterating that no substantial piece of writing is composed in a vacuum otherwise inhabited only by the author. There are many people I am glad to thank for their assistance and counsel throughout the process by which this book came together.

First and foremost, I couldn't have completed this project without the love and support of my partner, Woody Carrick, who has already heard far more about the Ashen Light from me than most people will ever care to know.

I'm indebted to my interview subjects, who answered my questions thoughtfully and patiently, and other experts who spent much time engaging with me on this topic. These include Prof. Paul Abel, Mary Stewart Adams, Dr. Dale Cruikshank, Dr. Andrew Crumey, Daniel Fischer, Dr. Candace Gray, Dr. William Hartmann, Dr. Vladimir Krasnopolsky, Dr. Richard McKim, Dr. Emilie Royer, Dr. Manuel Spitschan, and Prof. Christopher Russell. I'm grateful to the astrophotographers who kindly provided permission to reproduce their work here, namely Giuseppe Donatiello, Daniele Gasparri and John Sussenbach.

A few people saw substantially final versions of this manuscript before publication and offered comments and suggestions for improvements. I am particularly grateful to Francine Jackson (Brown University's Ladd Observatory) for making a very careful read and writing a very helpful review in repsonse.

The decision to embark on this project might not have been made without the advice of the British amateur astronomer and Ashen Light eyewitness, Richard Baum, whom I was fortunate to get to know right at the very end of his life. In my last correspondence with him in May 2017, just months before his death, Richard encouraged me to make the book "stimulating to thought," otherwise readers with pre-existing opinions about the Ashen Light would dismiss it, thinking there was nothing further about the subject to consider. "I have seen this only too often in my 87 years," Richard wrote me. "A saddening process in fact because they fail to realise the importance of what they pass over." I wish Richard lived long enough to see this book published, as I would have loved to know whether he felt I did justice to the topic.

I appreciate both that Springer took a chance on this topic and that Hannah Kaufman was my editor and Dinesh Vinayagam my production manager. Once again, it was a smooth and trouble-free experience bringing a title from concept to publication.

Lastly, thanks are due to innumerable friends, colleagues and members of my family who offered kind words and expressions of support throughout the project. I'm forever in your debt.

Tucson, AZ, USA
February 2021

John C. Barentine

Contents

Introduction

In the fading light minutes before sunset, I rack the telescope focus in and out before obtaining a sharp view of the Evening Star that has dominated twilight for the past several months. Each late afternoon, it is the first "star" visible in the deep blue of the western sky. I follow it for the next few hours, gradually lower toward the horizon, as the light of day fades and the brilliant crescent seems to shine more brightly. Next to other observations about it, I scrawl yet again in my notebook.

"Ashen Light not suspected."

It is, in fact, the same observation as every time I observed Venus during the last thirty years. The words are a reference to the appearance of faint light emanating from the night side of the planet, coming from a place where one has no expectation of seeing any light. It is pale and often devoid of color, or nearly so. It is the *lumière cendrée* of the French language, and the *Graulicht* of German. As ghostly a name as it is a spectre haunting the history of astronomy, in English it is the Ashen Light.

"Night side entirely dark."

Reports by other reputable observers are too numerous, and some of those observers too reputable, for me to completely dismiss the notion that *something* occasionally happens on the hemisphere of Venus facing away from the Sun that yields direct and very real light, the quiet rain of which falls partly into the telescopes of Earthbound observers, tickling their retinas in just the right way as to lead their brains to believe that they have seen something as real as the planet itself. Some argued passionately, to their very last days, for the objective certainty of what they saw. But others insisted with equal passion that they *never* saw it, even after a lifetime of dedicated Venus watching.

"Region away from the bright crescent examined for light. None seen."

Despite actively searching for most of my life, I can't say that the presence of the Ashen Light has ever suggested itself through the eyepiece of my telescope. Not once. And yet the mystery is so enticing that it became an obsession of four years' running by the time this book went to press.

My first encounter with the Ashen Light story came in the late 1980s when I was gifted a copy of James Muirden's *Amateur Astronomer's Handbook*, first published in 1974. Like nearly every other astronomy book I encountered as a kid, I read it from cover to cover because every aspect of astronomy seemed fascinating. On taking up the chapter on Venus, I came across a description of the Ashen Light that spanned only a few paragraphs, seemingly thrown in for completeness. Muirden introduced the Light with maybe the most succinct statement about it ever committed to print: "A phenomenon which has given rise to much dispute, even though its occurrence seems established by the weight of observation, is the occasional very faint luminosity of the dark side, aptly termed the *Ashen Light.*"

I'm still taken aback by those words. How could anything seen by visual observers using small telescopes for so long be controversial? By the time I read Muirden, humans had not only sent spacecraft to Venus, but they even managed to land a few on its hellish surface. It didn't make sense that something might be going on in the atmosphere of Venus that was powerful enough to produce light observable from Earth, and yet there was not so much as a single photograph that objectively demonstrated its existence. Muirden doesn't sound like much of a skeptic, deferring to the weight of centuries of reports by reliable observers. Instead, he probed at the edges of what might be a plausible physical explanation:

> What is the Ashen Light? We do not know and can therefore only theorize, but it seems possible that it could be caused by intense auroras in Venus' atmosphere. We must remember that Venus is relatively close to the sun, and so receives much more radiation than does the earth. If Ashen Light sightings could be tied in with solar activity, the evidence would be conclusive.[1]

Starting from a belief that the phenomenology of the Ashen Light is the first and best source of information that leads to informed speculation about

[1]Muirden, 167–168.

its cause, I started my own quest to understand the subject by collecting as many descriptions of sightings of the Light as possible. Two years of searching yielded nearly 500 individual published reports spanning some three centuries. Combing through those observations and noticing certain patterns and repeated themes was the genesis of this book, as was finding out (to my surprise) that no such work already existed.

The project led me to dip a toe into original research on the subject, to consider what other areas of observational astronomy remain incompletely explored by amateurs and professionals alike, and to gain (and in short order, to lose) a friend of unimpeachable expertise in this field, whose parting words to me were to "make [this book] popular and stimulating to thought; otherwise many will glance at it then walk away thinking they know it all."

At the end of this work, I come to a few broad conclusions about the Ashen Light and what it says about us as fully fallible human beings.

First, the human eye and brain are, as a system for detecting and processing the signals of very faint light, vastly underrated in their efficacy. By the first half of the twentieth century, as astronomy was overtaken by astrophysics as the more princely of the disciplines during the ramp-up of research spending fueled by the post-war economic boom, the photographic process displaced visual observations as the more reliable recording medium. At the same time, the push to construct ever-larger telescopes demanded technology that could wrest from the universe the secrets encoded in the steady arrival of cosmic photons on Earth, extracting every drop of information possible from every particle of light, some of which required billions of years to even reach our planet. Digital detectors, with much higher efficiency, in turn fully displaced the best photographic emulsions by the dawn of the new century. Visual observing was relegated to a pastime of amateur astronomers.

Second, the ways in which our senses couple to both memory and logic result in the firm belief in the objective reality of what those senses tell us about the world: at some level, seeing really *is* believing. It can be argued that visual impressions are too impermanent and too imprecise to be considered reliable, while imaging processes are the only means of achieving the objectivity that science demands. Yet to discard the careful records of eyewitnesses is to downplay billions of years of evolution by natural selection that has given humans tremendous sensory capabilities. Our eyes are sensitive to a dynamic contrast range of light comprising some twenty stops, or a ratio of a million to one. But they are coupled to brains, incredibly complex organs that also give us the credulity of superstition, a territoriality that sorts us into warring tribes, and the tendency to follow leaders blindly that has brought our species to the edge of obliteration and back in just the past century. Humans are often led to firm belief in the objective existence of things that very clearly never were, through observations of the world that left vivid memories and deep impressions. Our faith in the reality of our perceptions of the world, though devout, does not prevent us from being completely wrong.

Finally, while remaining formally agnostic about the existence of the Ashen Light—much less any specific explanation for it, if real—I believe the eyewitnesses throughout history who reported seeing it. I think they were reasonably convinced of the authenticity of what their eyes and brains told them about the information their telescopes collected when pointed toward Venus. At the same time, the prospect of a definitive explanation that will satisfy every skeptic seems dimmer than ever. Perhaps someone in the future will yet produce the unassailable evidence that either finally establishes or disproves the Ashen Light as a real, physical phenomenon. In either case, we would learn something important about human perception from simply knowing the right answer. And while this book certainly won't be the last word on the subject, I hope it is seen as helpful in collecting together in one place as much of the evidence for and against the Ashen Light as one author can.

So with that, a great story about the planet Venus begins with an equally great story about the planet Mars. And perhaps the greatest misapprehension in the history of astronomy began with an unfortunately bad translation of a single Italian word.

1

Prologue: The Martians That Never Were

"Little memory, no genius, much patience, and an everlasting curiosity about everything," was how Giovanni Virginio Schiaparelli (Fig. 1.1) once described himself,[1] but that self-deprecating review of his own intellectual capabilities belied his contribution to nineteenth-century astronomy. Although an observer possessed of keen perceptive abilities, he was also nearsighted and colorblind, crucial characteristics that likely influenced his judgment and left a particular imprint on how we now view the reliability of the human eye and brain in making useful astronomical observations.

Born on March 14, 1835, in Savigliano, an ancient city in what was then the Kingdom of Sardinia, Schiaparelli learned the lore of the night sky "as an infant," in his own words. His father, Luigi Schiaparelli, was a brick- and tile-maker from a long line of kilnmen, but he evidently believed in the value of education for his son, the first of eight children born to him and his wife, his third cousin Caterina Schiaparelli. Luigi instructed him in writing and mathematics, while Caterina taught him to read.

Through his father, Schiaparelli "came to know the Pleiades, the Little Wagon, the Great Wagon, and the *Via Lactea*." After seeing the trails of meteors blazing across the night sky, he asked Luigi what they were. "My father," he later wrote:

> answered that this was something the Creator alone knew. Thus arose a secret and confused feeling of immense and awesome things. Already then, as later, my

[1] 1907 letter to Professor Giovanni Marchesini, director of the Review of Philosophy, Pedagogy and Other Sciences, quoted in Mazzucato (2006).

J. C. Barentine, *Mystery of the Ashen Light of Venus*, Astronomers' Universe,
https://doi.org/10.1007/978-3-030-72715-4_1

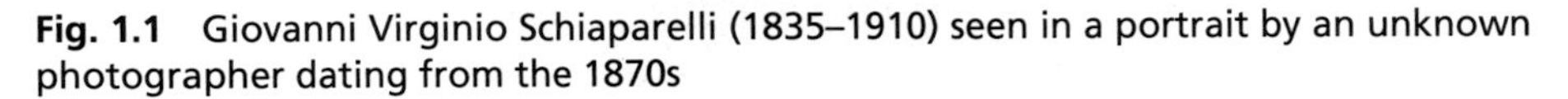

Fig. 1.1 Giovanni Virginio Schiaparelli (1835–1910) seen in a portrait by an unknown photographer dating from the 1870s

> imagination was strongly stirred by thoughts of vastness, of space as well as of time.[2]

He later cited as a crucial formative moment the total solar eclipse that swept across the Italian Piedmont on the morning of July 8, 1842. Writing to the journalist Onorato Roux toward the end of his life, Schiaparelli described how the event shaped an inquisitive view of his world:

> I put on my trousers quickly, I went to the window: it was just the time of total disappearance of the solar disc. … My wonder increased even more when I was told that some men were able to predict such phenomena by date and time. I had, then, the wish to be one of them and the ambition to witness the forces that govern the universe.

Under the pale glow of the solar corona, a young Schiaparelli "formed the ardent desire of participating in the counsels governing the universe."[3] He wanted a seat at the table among the arbiters of the heavens.

While still in Savigliano, he attracted the attention of Paolo Dovo, a learned man and priest at the church of Santa Maria della Pieve; later, Schiaparelli fondly remembered him as "a man of gold, a great lover of astronomy, and

---

[2]Bianucci, P. 1980. Giovanni Virginio Schiaparelli, *L'Astronomia*, 6, 45.

[3]1878, 'Osservazioni e fisiche sull'asse di rotazione a sulla topografia del pianeta Marte,' Memoria Prima, *Reale Accademia dei Lincei*; 1930, *Le Opere di G. V. Schiaparelli*, Milan, (reprint, New York, 1969), 1, 11–12.

one whose image could never be erased from the memory of those who had known him." Dovo gave Schiaparelli his first formal instruction in astronomy, lending him books and providing the young man with his first views of the night sky through a telescope fixed in the church's campanile. The magnificent universe expanded before him, the telescope revealing Saturn's rings, the phases of Venus, and the bright moons of Jupiter discovered two centuries earlier by his fellow countryman, Galileo Galilei.

His parents were sufficiently well-off such that they could afford to enroll him at the Gymnasium Lycée of Savigliano, which he began attending in the same year that his young eyes watched the shadow of the Moon darken the countryside of the Piedmont. In 1850 he enrolled at the University of Turin, completing a degree with distinction in hydraulic engineering and civil architecture. He was an excellent student; memorializing him in the pages of the *Monthly Notices of the Royal Astronomical Society* in the year after his death, the English amateur astronomer Edward Ball Knobel (1841–1930) wrote that in Turin Schiaparelli "gained a high reputation with his professors, and rapidly outdistanced his fellow-students in the study of pure and applied mathematics and in drawing."

Upon graduation, Schiaparelli remained for a while in Turin, teaching mathematics at the Gymnasium of Porta Nuova in order to earn a living. However, in Knobel's words, he found it "a position that was distasteful," so he petitioned the government of the Kingdom of Sardinia for financial support in order to further his astronomy studies abroad. His request was successful, and in 1857 he relocated to Berlin to study under Johann Franz Encke (1791–1865), most famous for calculating the orbital elements of the periodic comet that now bears his name. While in Berlin he also studied philosophy, meteorology, and geography, among other subjects.

In 1859, after a brief stint at Potsdam Observatory, he got a job at the Pulkovo Observatory near St. Petersburg, Russia, working for Friedrich Georg Wilhelm von Struve (1793–1864), his son Otto Wilhelm von Struve (1819–1905), and Friedrich August Theodor Winnecke (1835–1897). His performance at Pulkovo landed him a permanent position as second Astronomer at the Brera Observatory at Milan under Francesco Carlini (1783–1862). Carlini's death shortly thereafter, and Schiaparelli's keen intellect and scientific output, led to his appointment as Director. He remained in Milan for the rest of his life, making his permanent professional home at Brera.

By the 1870s, Giovanni Schiaparelli had become one of the world's pre-eminent astronomers. Using a 4-inch telescope, he discovered the asteroid (68) Hesperia in 1861, naming it for the ancient Greek word for Italy. In the mid-1860s, comparing the orbits of various comets to the directions on the sky from which certain meteor showers appear to radiate, he correctly deduced the causal relationship between the streams of dusty debris shed from comets and swarms of "shooting stars" whose timely reappearance each year had puzzled observers for millennia. He further contributed a quarter-century of precise measurements of the positions of double stars and made extensive observations of Mercury and Venus, venturing (incorrect) estimates of their rotation periods.

While "his eye was famous for its keenness," wrote the clinical psychologist and historian of astronomy, William Sheehan, it was "not without its peculiarities." Schiaparelli was "severely myopic" and he suffered from red-green color blindness, a fact that certainly affected his perception of subtle shading on the discs of the planets he saw through the telescope eyepiece. Although he acknowledged privately that his eyes were "only slightly sensitive to the nuances of colour,"[4] Schiaparelli still achieved fame as one of the most distinguished visual observers of his generation.

Although in existence for a century by the time Schiaparelli arrived there, Brera Observatory was far from a cutting-edge research center. Its instruments were dated and inadequate for pursuing his research agenda. The nascent Kingdom of Italy took note of his discoveries, and the Minister of Public Education appropriated funds to provide the observatory with an equipment upgrade. In February 1875, Schiaparelli oversaw the installation of a brand-new refracting telescope, its 22-centimeter objective lens figured by the celebrated Bavarian lens maker Georg Merz (1793–1867). Although on the small side of the range of instruments available to observers at the world's premiere research facilities, the new telescope placed within Schiaparelli's reach the power to make important contributions to planetary astronomy.

It was through the Merz refractor that Schiaparelli began observing Mars during the dog days of the summer of 1877. That he spent some time on Mars was practically incidental to what he considered the important work on Venus and Mercury, though both planets suffered from the circumstance that they never appeared very far from the Sun in the sky and hence were invisible for most of the night. The Red Planet made for a useful point of comparison, allowing Schiaparelli to gauge the quality of the new telescope. While he did not intend then to dedicate himself to a "protracted series of regular observations," he wrote that he:

[4]1882, Proceedings of the Meeting of the Royal Astronomical Society, *The Observatory*, 5, 135–137.

> desired only to experiment to see whether our refractor of Merz, which had given such good performance on double stars, possessed the necessary optical qualities to allow also for the study of the surfaces of the planets. I desired also to verify for myself what the books of descriptive astronomy expounded about the surface of Mars, its spots and its atmosphere. I must confess that, on comparing what I saw on the planet with the maps that had been most recently published, my first attempt did not seem very encouraging.[5]

Nevertheless, he persisted in his Mars work, and his drawings proved to be as good as the best then-published.

In the latter half of 1877, Mars was situated nearly opposite the Sun in the night sky and available for viewing all night long. Schiaparelli undertook a careful examination of Mars at this "opposition," having become convinced that useful work was to be done delineating its topography. At the time, Schiaparelli shared the prevailing opinion of the planet among astronomers, formed in the days of the earliest telescopic views in the seventeenth century: Mars was a watery world where bright areas on its surface indicated landmasses floating in the voids of dark seas. These, however, were always shadowy and indistinct to earlier observers, and Schiaparelli felt that existing maps of the planet were insufficient to draw meaningful conclusions about the true nature of its surface.

The astronomical "seeing" from his rooftop observatory in the autumn of 1877 was remarkably good, to the extent that in October he experienced certain moments of near-perfect atmospheric calm in which entirely new sights unfolded to him through the eyepiece of the Merz refractor. And in those interludes when the image of Mars steadied amidst the roiling vapors above Milan:

> it seemed as if a dense veil were removed from the surface of the planet, which appeared like a complex embroidery of many tints. But such was the minuteness of these details, and so short the duration of this state of affairs, that it was impossible to form a stable and sure impression of the thin lines and minute spots revealed.[6]

A new planetary world was thusly born, the nomenclature of its features inspired by the mythology of classical antiquity. As Sheehan tells the story,

---

[5]1878, "Osservazioni e fisiche sull'asse di rotazione a sulla topografia del pianeta Marte," Memoria Prima, *Reale Accademia dei Lincei*; *Opere*, 1, 61.

[6]1889; "Ueber die Beobachtungen Erscheinungen auf der Oberflache des Planeten Mars", *Himmel und Erde*, Vol. 1, Berlin; also in *Opere*, 2, 23.

Schiaparelli "introduced names which have ever since been part and parcel of the romantic lore of the red planet: Syrtis Major, Sabaeus Sinus, Margaritifer Sinus, Solis Lacus, Juventae Fons, Hellas, Elysium, Tharsis." Among the new terms to describe and label what he saw on the Martian surface, one stood out fatefully.

Canali.

Through the eye of the lens, indistinctly in the beginning and then with greater certainty, Schiaparelli gradually convinced himself of the reality of various streaks and shadings that appeared to him with striking linearity. In August he identified the first such feature, naming it Ganges after the great Asian river with its vast delta, tracing the drainage of northeastern India from a wide channel into an array of rivulets emptying into the Bay of Bengal. He saw through his telescope the geography of water, and he used the corresponding Italian words to describe what he felt must be the courses of waterways. *Canale* (meaning, variously, "channel," "duct," or "gully") and *fiume* ("river") appear interchangeably in his writing to denote many of the delicate lines he saw on the disc of Mars. But there is some ambiguity in how the word *canale* might be rendered, given that it also carries connotations of its English cognate, *canal*, which is readily distinguishable as a deliberate modification of the land. Schiaparelli's words, translated into English, carried much more of the artificial sense than it seems he ever intended, even though he didn't protest too loudly against the interpretation his words received on the other side of the Atlantic.

It turns out that Schiaparelli's use of *canali* with reference to Mars wasn't new. The term was used as early as the 1850s by Father Angelo Secchi, S.J. (1818–1878), Director of the Observatory at the Pontifical Gregorian University. In 1858, Secchi identified perhaps the largest and most noticeable of the dark regions on Mars as the "Atlantic Canale," but it is Schiaparelli's name – Syrtis Major – that is found on maps of the Red Planet to this day.

Schiaparelli's view was initially somewhat skeptical, and while he seems to have thought that Ganges was not a real river delta, he acknowledged that it was certainly a unique landform. But as late summer progressed into autumn, one by one, linear features seemed to unmistakably appear to him everywhere:

> In most cases the presence of a *canale* is first detected in a very vague and indeterminate manner, as a light shading which extends over the surface. This state of affairs is hard to describe exactly, because we are concerned with the limit between visibility and invisibility. Sometimes it seems that the shadings are mere reinforcements of the reddish colour which dominates the continents — reinforcements which are at first of low intensity. … At other times, the appearance may be more that of a grey, shaded band… It was in one or other

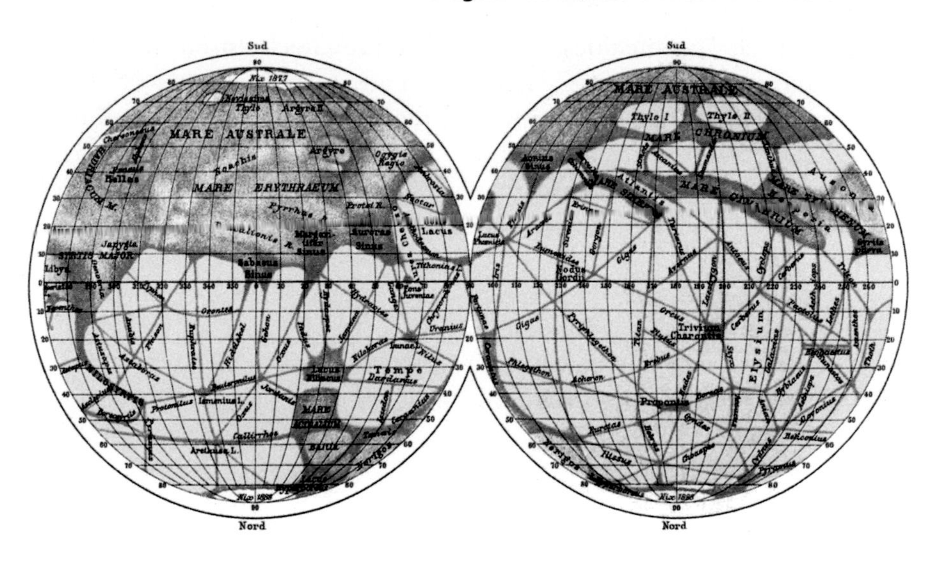

Fig. 1.2 "General map of the planet Mars according to the observations made in Milan from 1877 to the present" from Giovanni Schiaparelli's *La Vita Sul Pianeta Marte* (1893). The map's two hemispheres represent a synthesis of nearly two decades of Schiaparelli's visual observations of Mars

> of these indeterminate forms that, in 1877, I began to recognise the existence of the Phison (October 4), Ambrosia (September 22), Cyclops (September 15), Enostos (October 20), and many more.[7]

In the same year that Father Secchi died, Schiaparelli published his first map of Mars showing the *canali* as well as other shadowy features he patiently observed through the Merz refractor at Brera. The map evolved and was refined over the following two decades; a summary published in 1893's *La Vita Sul Pianeta Marte* ("Life on Planet Mars") is shown in Fig. 1.2. His imperfect eyes saw Mars dissolve into a tangle of lines, stretching predominately across the planet's southern hemisphere. They seemed to bridge the broader dark patches, as though situated in such a way as to connect together great bodies of water by a vast network of straight, narrow earthworks. Although Schiaparelli didn't say as much, the implication of his maps was hardly disguised: Mars was covered in an elaborate system for the intentional geographic redistribution of water. The *canali* weren't simply natural waterways. They were engineered rivers.

To further complicate matters, after the Mars opposition of 1881–1882 Schiaparelli began drawing nearly all the *canali* as sets of doubled lines, which

[7] *Opere*, 1, 164.

he referred to as "geminations," a term that the Greek astronomer Eugène Michel Antoniadi (1870–1944) later said "made a Sphinx-world of Mars."[8] Although Schiaparelli himself seemed a little taken aback at what he thought he saw during that Martian apparition in a handful of instances, he gradually came to identify geminations in virtually every linear feature on Mars that he perceived. Knobel wrote in Schiaparelli's obituary that the Italian astronomer:

> was inclined to think that the gemination of the lines was periodic, and made its appearance when the heliocentric longitude of Mars was about 110° or 120°, some 2 months after the spring of the northern hemisphere. Such unexpected phenomena excited a large amount of criticism and scepticism, which indeed up to the present day is not yet allayed.[9]

Was Schiaparelli simply predisposed to see linear features on other planets through his telescope? His work on the planet Mercury gives some insight. In his extensive observations of the innermost planet, he thought he saw shadings indicative of the planet's true surface, enabling him to make an estimate of the planet's rotation period. Due to an alignment between Mercury's "solar day" (the time between successive sunrises at any location on its surface) of 176 Earth days and the time between successive morning or evening elongations of the planet, 117 Earth days, observers see alternating faces of the planet at each elongation east or west of the Sun. From this, assuming the same telescopic appearance in each case, Schiaparelli and others concluded incorrectly that the rotation period of Mercury must equal its orbital period about the Sun, 88 days. In fact, it was not until nearly a half-century after his death that radio astronomers determined a very precise—and much shorter—rotation period of 58.65 days.

Nevertheless, Schiaparelli saw what he saw, and in 1889 he noted his impression, "when the seeing with the instrument becomes very steady, that all the appearances are resolved into very fine formations."[10] His certainty in the objective existence of the lines crystallized during the previous year's Martian opposition. Quoting him nearly two decades later, the American zoologist and orientalist Edward Sylvester Morse (1838–1925) wrote that Schiaparelli declared:

[8] 1898, *Memoirs of the British Astronomical Association*, 6, 102.

[9] Knobel, 284.

[10] 'Sulla Rotazione e Sulla Costituzione del Pianeta Mercurio', in *Opere*, 5, 333–343.

> "the *canali* had all the distinctness of an engraving on steel, with the magical beauty of a colored engraving." He furthermore says: "As far as we have been able to observe them hitherto, they are certainly fixed configurations upon the planet, the Nilosyrtis has been seen in that place for nearly 100 years and the others for at least 30 years."[11]

Yet Schiaparelli admitted that the lines revealed themselves only in moments of particular atmospheric calm, while in other instances they became blurred and indistinct as the air tumbled about turbulently over the telescope dome. While he may well have believed the features he saw to be so narrow such that they only truly became visible in brief glimpses when the air settled down, the circumstances allow for some degree of "postselective discretion," as Sheehan put it, enabling the interpretation of observations "to fit one's scheme."

Other astronomers took note of Schiaparelli's claims and began turning their own telescopes toward the Red Planet. What followed may be the best-known case of a kind of mass delusion recorded in the annals of the history of astronomy. One by one, astronomers throughout Europe and the United States began to report confirmations of Schiaparelli's *canali* through various telescopes and under all manner of observing conditions. Even the narrowness of the linear features he reported, attributable to the optical phenomenon of diffraction and the more limited resolution of the smaller telescope he used in comparison with those deployed by other observers, was noted and carefully recorded by others. Because Professor Schiaparelli said he saw straight lines on Mars, others reasoned that they must exist.

Awareness and eventual acceptance of the geminations waited for several years to pass. In 1886 two French astronomers, Henri Joseph Anastase Perrotin (1845–1904) and Louis Thollon (1829–1887), confirmed the appearance of the doubled lines through the Nice Observatory's 74-centimeter refracting telescope. Schiaparelli was ecstatic on hearing the news. "I attach very great importance to this confirmation for people will hereafter cease to scoff at me in certain places," he wrote. "The geminations are very difficult to explain, but it is indeed necessary to admit their existence."[12] Two years later, Perrotin muddied the waters by publicly announcing that the "continent" of Libya had disappeared since the previous opposition, declaring that it "no longer exists today." Results from other observatories, on the other hand, found no changes to Libya.

---

[11] 1906, *Mars and its Mystery*, Boston: Little, Brown & Co., 59.

[12] *Corrispondenza*, 1, 153.

Some observers affirmed the appearance of the canals but without the geminations, while others saw no canals at all. There were objections both to the existence of the straight lines and the interpretation that they must represent some kind of artificial landscape engineering effort. Nathaniel Everett Green (1823–1899), an English professional painter and astronomer, complained during an 1890 meeting of the British Astronomical Association that Schiaparelli and other proponents of the *canali*:

> have seen them, so that they must be there. That other observers have seen whatever forms the basis of these lines I do not for a moment doubt, but I feel thoroughly convinced they *have* not drawn what they have *seen*, or, in other words, have turned soft and indefinite pieces of shading into clear, sharp lines.[13]

Schiaparelli remained coy on the subject for more than a decade after publishing the results on Mars, neither fully and publicly acknowledging nor endorsing the implication of some kind of intelligence probably responsible for creating his *canali* until 1893. His openness toward the possibility of a visible manifestation of intelligent life on Mars attracted the interest of some of the greatest names of nineteenth-century astronomy, including the Frenchman Camille Flammarion (1842–1925), who illustrated the canals in 1894's *Le Terres du Ciel* (Fig. 1.3), and the American businessman Percival Lowell (1855–1916), who at the same time as Schiaparelli's admission about life on Mars was building his own great observatory on the aptly named Mars Hill just outside the town of Flagstaff, Arizona.

In his sensational *Mars and Its Canals* (1906), Lowell made clear that he bought the theory lock, stock, and barrel: "The strange geometricism which proves inexplicable on any other hypothesis now shows itself of the essence of the solution," Lowell wrote. It all fit together beautifully; the broad dark spots he himself observed with his 24-inch Alvan Clark and Sons refractor at Flagstaff were Martian oases and "clearly ganglia to which the canals play the part of nerves."[14] These featured prominently in virtually all of his drawings (Fig. 1.4) from the establishment of Lowell Observatory to his death.

In the span between Schiaparelli's description of the *canali* and the dawn of the Space Age, the story took on elaborate and dramatic details: the canals were dug in desperation by a dying Martian race to conduct water from the planet's polar ice caps to the parched desert cities of its equatorial regions as a last-

[13] 1890, Report of the meeting of the association held December 31, 1890, *Journal of the British Astronomical Association*, 1, 112; emphasis in the original.

[14] 1906, *Mars And Its Canals*, London: MacMillan & Co., 365.

Fig. 1.3 "Le lever du soleil sur les canaux de Mars" ("Sunrise over the Canals of Mars") in Figure 31 from Camille Flammarion's *Les Terres Du Ciel* (1884). The romantic notion of watery Martian landscapes persisted in popular works of science well into the twentieth century

ditch means of saving the Martians from certain extinction. In this hypothesis was a parallel that followed the development of technology from the promise of industrialization's nineteenth-century march to the devastation of World War I in Europe, wrought by twentieth-century technology meeting perhaps humanity's oldest impulse—to rain destruction down upon itself. The Martian canals were, therefore, both an allegory and a cautionary tale. No matter the details, what they implied was clear: humanity was not alone in the cosmos, and intelligent life had arisen essentially right next door to our own world.

It was enough for the *New York Times* to scream rather startlingly **THERE IS LIFE ON THE PLANET MARS** in a December 9, 1906, headline

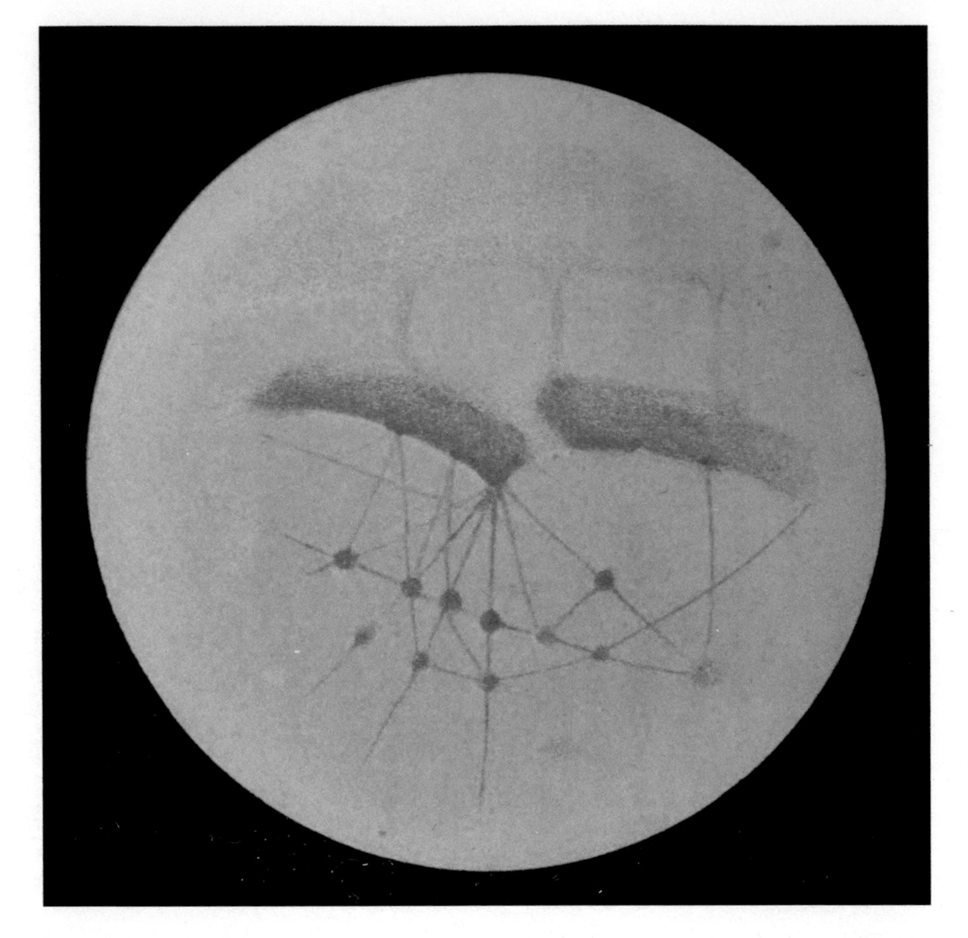

Fig. 1.4 "Mars. Sinus Titanum. November, 1894." Plate 1 from Percival Lowell's *Mars* (1895)

(Fig. 1.5). The story's author, Emily Lilian Whiting (1847–1942), breathlessly asserted that not only Schiaparelli's canals, but also the manifest existence of their intelligent makers, were incontrovertible facts determined after sufficient scientific scrutiny, debate, and resolution:

> The hypothesis of canals on Mars has already emerged from its progress through the usual stages of skepticism, ridicule, and denial which every new advance in science has to encounter. It seems to be the law and the prophets regarding all phases of the conquering of the unknown. "Every generation," remarks Mrs. Julia

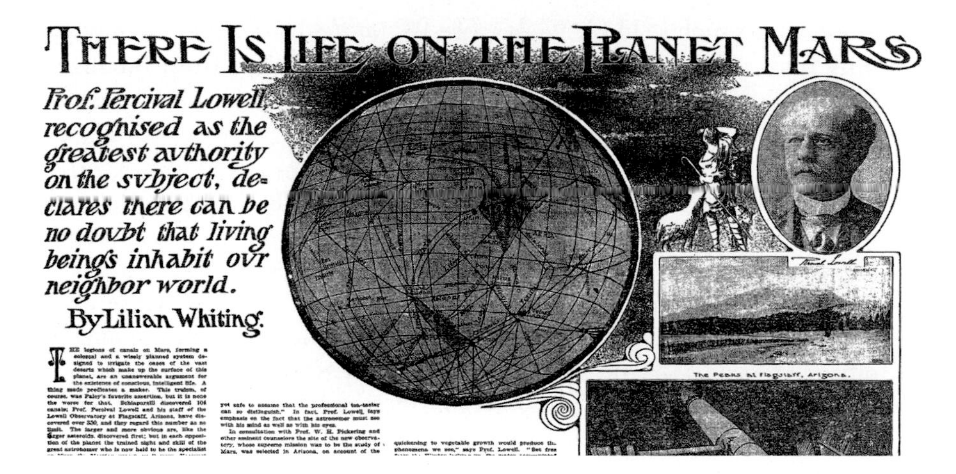

THERE IS LIFE ON THE PLANET MARS

Prof. Percival Lowell, recognised as the greatest authority on the subject, declares there can be no doubt that living beings inhabit our neighbor world.

By Lilian Whiting.

THE legions of canals on Mars, forming a colossal and a wisely planned system designed to irrigate the oases of the vast deserts which make up the surface of this planet, are an unanswerable argument for the existence of conscious, intelligent life. A thing made predicates a maker. This truism, of course, was Paley's favorite assertion, but it is none the worse for that. Schiaparelli discovered 104 canals; Prof. Percival Lowell and his staff of the Lowell Observatory at Flagstaff, Arizona, have discovered over 330, and they regard this number as no limit. The larger and more obvious are, like the larger asteroids, discovered first; but in each opposition of the planet the trained sight and skill of the great astronomer who is now held to be the specialist

yet safe to assume that the professional tea-taster can so distinguish." In fact, Prof. Lowell lays emphasis on the fact that the astronomer must see with his mind as well as with his eyes.

In consultation with Prof. W. H. Pickering and other eminent counselors the site of the new observatory, whose supreme mission was to be the study of Mars, was selected in Arizona, on account of the

quickening to vegetable growth would produce the phenomena we see," says Prof. Lowell. "Set free

Fig. 1.5 *New York Times* headline of December 9, 1906, trumpeting Percival Lowell's certainty that the Red Planet was inhabited by an intelligent species

> Ward Howe,[15] "makes a fool of the one that went before it," and, as Prof. Lowell says, "the generation that witnesses an advance is not the generation to believe in it." But a persistent incredulity may be no less stupid and dense than simple incredulity. Unreasoning denial has no more claim than unreasoning acceptance. That there are canals on Mars is to-day an accepted fact of astronomy, by any doubt of which the doubter merely argues himself unknown. The establishment of this fact is so recent as to be one of the most important truths of the new astronomy which lies well within the past half century.

"The only logical result that can be reached from these fundamental, these demonstrated and demonstrable facts," Whiting concluded, is that the existence of the Martian canals "is an unanswerable and an absolute proof that there is conscious, intelligent, organic life on Mars."

Science fiction novelists exploited this conclusion to great effect, from Edgar Rice Burroughs's *Barsoom* series (1912–1943) to Ray Bradbury's *The Martian Chronicles* (1950), and contemporary authors still sometimes invoke Martian canals as a background against which the planet is terraformed by humans.[16] Others capitalized on Victorian anxieties over the threat of foreign military

[15] Howe (1819–1910) was an American author and social activist who advocated for both women's suffrage and the abolition of slavery. As a poet, she is best remembered for writing "The Battle Hymn of the Republic" (1861).

[16] See, e.g., Colin Greenland's *Take Back Plenty* (1990); Kim Stanley Robinson's *Mars* trilogy (1993–1999) and *2312* (2012); and S. M. Stirling's *In the Courts of the Crimson Kings* (2008).

invasion in works such as H.G. Wells's *The War of the Worlds* (1897), in which humanity finds itself on the other end of the telescope lens from intelligent Martians with their eyes on our resources:

> No one would have believed in the last years of the nineteenth century that this world was being watched keenly and closely by intelligences greater than man's and yet as mortal as his own; that as men busied themselves about their various concerns they were scrutinised and studied, perhaps almost as narrowly as a man with a microscope might scrutinise the transient creatures that swarm and multiply in a drop of water. … At most terrestrial men fancied there might be other men upon Mars, perhaps inferior to themselves and ready to welcome a missionary enterprise. Yet across the gulf of space, minds that are to our minds as ours are to those of the beasts that perish, intellects vast and cool and unsympathetic, regarded this earth with envious eyes, and slowly and surely drew their plans against us.[17]

As for Schiaparelli, his eyesight seems to have deteriorated through the last two decades of the nineteenth century. Toward the end of his Martian observations, he confided to Antoniadi that "during 1894 (or 1896) … the field of vision was becoming darker and darker. At last, in 1900, I realised, with great sorrow, that the images were also becoming deformed. Then I made my farewell."[18] He decided not to publish any of his Mars observations made after 1890.

Schiaparelli received many awards for his astronomical work, was elected to various scientific societies, and even was appointed a senator of the Kingdom of Italy in 1889 for his services to the nation. He retired from Brera Observatory in 1900, at which point he took up other academic pursuits in order to keep his mind active before dying of a cerebral blood clot on July 4, 1910. At his death, the President of the Italian Council of Ministers, Luigi Luzzatti, wrote in a personal telegram to the Schiaparelli family that "a ray of celestial thought has died: Italy lost its greatest and most glorious scientist." At the time of his death, it was recognized that "Schiaparelli had proved himself in so many ways to be a most careful, accurate, and conscientious observer, that though some observers of high excellence have altogether failed to confirm his observations of these 'canali,' it is impossible to believe that they have not some objective existence."[19]

[17] 1898, *The War of the Worlds*, London: Chapman and Hall, 3–4.

[18] Letter of August 29, 1909, quoted in Antoniadi, E. M., 1930, *The Planet Mars*; translated by Patrick Moore, Shaldon, 1975, Devon: Keith Reid Limited, 45.

[19] Knobel, 284.

But it was all dead wrong. The perception of the *canali* as designed and constructed earthworks persisted nearly to the arrival of the first spacecraft from Earth in the 1960s, which showed in their photographs not the slightest trace of any identifiably artificial structure. Even though the notion of an advanced race of beings on Mars lost popularity during the interwar period as scientific discoveries about the planet showed it to host an environment seemingly inhospitable to life, some adherents held fast to the conventional wisdom. Others reverted to earlier ideas about the apparent waxing and waning of the dark spots on Mars with the seasons. These, they reasoned, might still be attributable to some simple, hardy, and decidedly non-intelligent organism.

The first advanced spacecraft to soft-land on Mars of the American *Viking* program of the 1970s tested the soil at two sites for evidence that something behaved under laboratory conditions like life as we know it, metabolizing nutrients and expelling waste products. Seemingly positive results at first generated excitement among researchers and a favorable media buzz, but further analysis showed that the apparent signs of life were most likely due to the chemistry of clays in the Martian soil whose structures were altered by eons of exposure to otherwise sterilizing levels of ultraviolet light from the Sun. The jury remains out on the question of whether Mars now hosts life or ever did, and researchers continue to look for clues that simple life forms might (still) exist on Mars under rare and highly favorable conditions.

The canals as artificial constructs, and their implied creators, never existed in the first place.[20] Yet in some sense, as William Sheehan wrote, "there is no question that Lowell was right about the artificiality of the canals of Mars. The canals were, however, artificial not because the Martians made them but because what we perceive is itself in an important sense an artifact."[21] Writing for his *Sky and Telescope* magazine blog, Tony Flanders argued that Lowell's "sin was lack of objectivity." Knowing fully well that the canals were controversial, Lowell was nevertheless "hell-bent on proving their existence. With an attitude like that, he could hardly help seeing them, despite the fact that they don't really exist."[22]

The Martian canal affair begs certain questions: what controls our perception of the universe around us? And how does it sometimes lead the brain

---

[20] Despite this conclusion, Schiaparelli was somewhat vindicated by spacecraft photographs that showed roughly linear, but entirely naturally occurring, features on some of the same parts of Mars for which he drew canali on his maps. q.v. Hartmann, W. K. 2003. *A Traveler's Guide to Mars*, New York: Workman Publishing, 44–49.

[21] *Planets And Perception*, 276.

[22] "The Reliability of Visual Observing," November 30, 2007. https://www.skyandtelescope.com/astronomy-blogs/the-reliability-of-visual-observing/.

astray? Certain aspects of the process of making an astronomical observation are independent of the observer and subject only to the laws of physics. Once decisions are made concerning the details of optical equipment used to collect cosmic light and bring it to a focus in the vicinity of the human eye, the die is to some extent cast. The diffraction of light, a property intrinsic to the wave aspect of photons, establishes a lower limit in terms of the power of a telescope to resolve fine details on a planetary surface. In particular, diffraction determines the smallest angle between two objects at which they can be just distinguished as separate. This corresponds to a physical length of separation if the distance to the object is known.

This would all be fine and well if the Earth were an airless world, although no doubt it would be inconvenient for its air-breathing inhabitants. However, our home is surrounded by a thick atmosphere rising almost 500 km high over our heads. The densest part of that atmosphere is mercifully located nearest us, making up the troposphere that extends only up to around 16 km. But it is in that first few kilometers where all of the relevant action in this story takes place, for a thick planetary atmosphere is subject to the complexities of fluid flow and transport. Light from space traversing this region suffers path distortions as it encounters more- or less-dense regions set to roiling by differences in temperature. Light rays are bent in new directions throughout their flight from the top of the atmosphere to the ground; some photons are scattered away at large angles from interactions with molecules or small dust particles, while others are absorbed completely.

Due to the chaotic nature of turbulent flows in the lower atmosphere, precise modeling of this behavior is almost impossible, and all we get is something like a statistical average of motions over whatever time interval we consider. Technologies like adaptive optics can help by correcting for atmospheric distortions of light rays in real time, but the inherent optical limitations of telescopes are hardwired into them. The story is already written, partly but indelibly, before the human eye and brain even have a look at the incoming torrent of visual information.

Even after the introduction of photography to astronomy in the mid-nineteenth century, firsthand visual observations through telescopes remained prized for their value. While the photographic plate effectively displaced the eye and brain within a 100 years of its introduction, the bigger story is the emergence during the same period of the modern science of astrophysics as the art of astronomy was increasingly relegated to the province of hobbyists. Photography of Mars using Earth-based telescopes as a means of resolving the controversy of the Red Planet's supposed canals was doomed to failure, in part

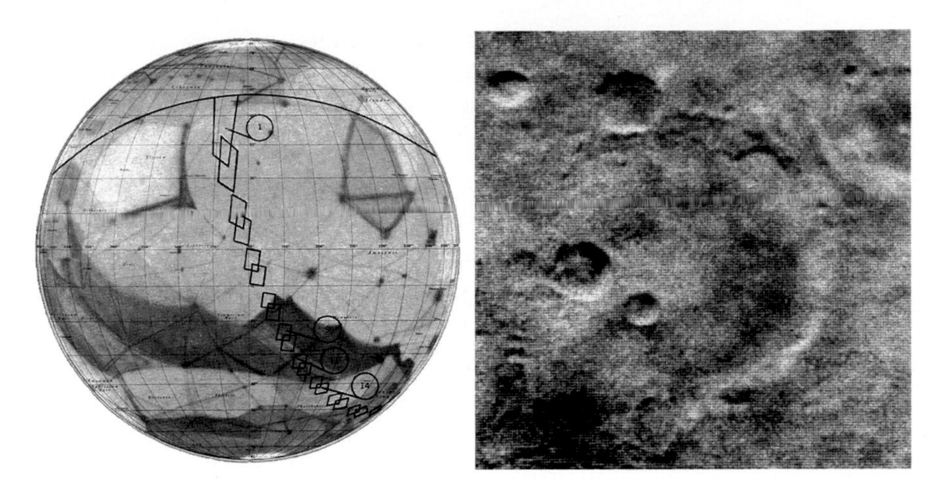

Fig. 1.6 *Left*: The imaging footprint of the *Mariner 4* spacecraft superimposed on a pre-Space Age map of Mars. A series of black diamonds traces the spacecraft's ground path as it flew past the Red Planet. *Right*: Frame 11E, the highest-resolution image of Mars returned by *Mariner 4* near its closest approach on July 15, 1965. It shows the eponymous 150 km Mariner crater at the edge of Sirenum Fossae (35 °S, 164 °W). The region shown is about 250 km on a side, and north is up. Both images courtesy National Aeronautics and Space Administration and the Jet Propulsion Laboratory

because it could not overcome the same optical and atmospheric factors that limited visual observations.

By the time imaging technology developed to be able to compensate for this shortcoming, Mars had already been visited by robotic spacecraft from Earth that provided photos of its surface with a resolution that will never be equaled from the ground. Humanity's first detailed view of the planet (Fig. 1.6) came in July 1965 when the American *Mariner 4* spacecraft flew past some 10,000 km above the Martian surface. The images showed no artificial earthworks and no evidence of a dying alien race desperately struggling to slake the thirst of its cities.

"Visual observation is superior to photography in the resolution of fine detail," wrote the Dutch-American planetary scientist Gerard Kuiper in 1955, "as is true in the observation of double stars. The reasons for this superiority have often been debated; my impression is that it is due in part to the fact that the eye 'works in series' with the brain and in part to the remarkable qualities of the eye as a recording instrument."[23]

[23] *Publications of the Astronomical Society of the Pacific*, 67(398), 271.

In the case of perceived linear features on Mars, perhaps Schiaparelli's brain dominated the series. Subconsciously or otherwise, perception overrode skepticism, and the brain delivered a message to the eye that returned in a kind of feedback loop. Schiaparelli's professional reputation may have, in turn, instigated a sort of similar feedback loop that, like a contagion, insidiously infected the brains of other observers, leading to a kind of mild hysteria in which those others felt their skills inadequate if they *didn't* also see what Schiaparelli insisted that he saw. But in the end, no amount of prestige or self-convinced certainty can offset the fact that some scientific observations–and the ideas developed to explain them–are simply wrong.

There had to be a first observation that touched off the firestorm of controversy around the Martian canals, for no one had previously reported anything quite like the "geminated" lines of Schiaparelli's Mars. What caught his eye in the autumn of 1877 was the spark that ignited that fire. But other planets, too, had their Schiaparelli moments, including Earth's nearest planetary neighbor. And for Venus, that moment came on a cold Italian night in the winter of 1643.

2

"I Could See the Dark Part of Venus..."

The "canals" of Mars are now viewed as an artifact of the era before the robotic exploration of the Solar System, all evidence for their reality, independent of the eyes and brains of the telescopic observers who recorded them, having vanished. And yet in at least some instances, there are hints that distinctly natural linear features on Mars inspired some of these observations. Other phantoms of the eyepiece, some of which we'll explore later, have similarly come and gone throughout astronomical history, supported at times with evidence yet still subject to a little too much credulity. But one remains some four centuries after it was first described, stubbornly refusing to either submit fully to scientific explanation or go away altogether.

After hundreds of years of scientific study, we confidently know a lot about the physical nature of Venus. Some five decades on since the first robot emissaries from Earth flew past it, we have characterized the planet in remarkable detail. Despite its nearly identical size, Venus bears almost no resemblance to our planet. Its dense atmosphere, composed almost entirely of carbon dioxide, traps heat from the Sun, raising the surface temperature above that of a household oven and a pressure equivalent to that found almost a thousand meters below the surface of Earth's oceans. In the middle layers of the atmosphere, winds reach speeds comparable to those of cruising commercial airliners on Earth. Opaque clouds of sulfuric acid droplets absorb and scatter essentially all optical light, rendering the surface of Venus permanently invisible from above (Fig. 2.1, left). Its large iron core seems thermally inert, driving neither plate tectonics nor an appreciable magnetic field. The planet spins once on its axis every 243 days—the slowest rotation period of any major

J. C. Barentine, *Mystery of the Ashen Light of Venus*, Astronomers' Universe,
https://doi.org/10.1007/978-3-030-72715-4_2

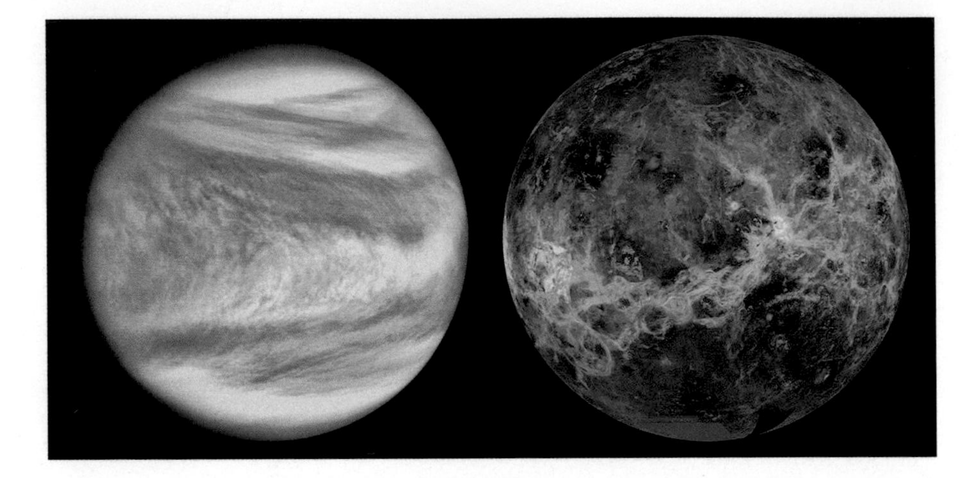

Fig. 2.1 Two spacecraft views of Venus. Left: NASA image ARC-1981-A78-9167 made in ultraviolet light by the *Pioneer Venus Orbiter* on September 29, 1981. The image is rendered in a colormap approximating the true-color appearance of the planet in visible light. Right: a global synthetic aperture radar view of Venus created from *Magellan* and *Pioneer Venus Orbiter* spacecraft data obtained in 1991, showing surface features otherwise obscured from view by its thick atmosphere. The false color suggests a visual impression of the surface illuminated by sunlight attenuated by its passage through the clouds. Both images courtesy of NASA/Jet Propulsion Laboratory

planet—in a direction opposite the sense of its orbit around the Sun. Radar measurements from orbit reveal a surface of many volcanoes and relatively few impact craters (Fig. 2.1, right). Murky images of its surface taken by Soviet spacecraft that soft landed on Venus in the 1970s and 1980s show low outcrops of platy, igneous rocks amid grayish basaltic soils. Scenes are illuminated by a dull, reddish sky light of approximately the same brightness as an overcast day on Earth.

In short, the surface and lower atmosphere of Venus are a boiling, acidic hell, while its insides appear geologically dead. Whatever common properties our worlds shared early in their respective histories, science has shown that they long since parted company and evolved along highly divergent paths.

Yet Venus beckons to us in an odd and enduring way. Like Mars, it has a lengthy presence in human consciousness that reaches back far into prehistory, occupying a seat deep in the recesses of time. It is the planet that seems to appear earliest in our cultural history, with mentions in some of the world's oldest written folklore. In turn, our distant ancestors' attention may well have been drawn to Venus at the start of our recorded history as humans made the important transition from nomads to farmers and began an intellectual journey that may one day take us to the stars. Like the Earth's Moon, Venus was an early cultural milepost along that route.

Venus in Myth and Tradition

To the ancients in the West, Venus was a conspicuous "wanderer" in the night sky, chief in brightness among starlike objects, always feminine in identity, defined by pairs of attributes reflecting various human motivations. The recognition of Venus as separate from both stars and the other planets was an important step in the expansion of human consciousness to recognize the existence of a space beyond our own Earth. Nearly 6,000 years ago, during the millennium between the protohistoric Chalcolithic period and the Bronze Age, the earliest settled communities appeared in Mesopotamia. Atop a hill above what is now a dried-up, ancient channel of the Euphrates River in present-day southern Iraq, a city called Uruk emerged in phases beginning around 4000 BCE. From nearly its founding, Uruk was strongly associated with a goddess called Inanna, whose name derives from a Sumerian phrase meaning "Lady of Heaven." To the Akkadians, Babylonians, and Assyrians, she was Ishtar; to the Phoenicians, she was Astarte. As a deity, she had it all: in the mythology of Sumeria, she was the goddess of love and lust, beauty and fertility, knowledge and wisdom, and war and combat.

In addition to being one of the most widely venerated figures in the Mesopotamian pantheon, Inanna was identified with the planet Venus. It comes as little surprise that Sumerians associated the planet with oscillations between pairs of attributes, given the careful observations of the sky, day, and night, on which they relied to keep an accurate calendar for coordinating activities attendant to agriculture. Over many months, the planet moves away from the Sun in the sky, reaching a maximum separation before moving back toward the Sun and disappearing for several days before its apparent rebirth. Venus became a symbol of recurring patterns in nature, both human and divine.

From the regular apparition, and similar appearance, of both a "morning star" and an "evening star," Sumerian astronomers concluded that the two were manifestations of the same thing. That duality may well have inspired a tendency to project onto Venus all manner of dichotomies, Inanna's vacillation between aspects reflecting some knowledge of a heavenly body that never strayed far from the Sun. As the next-brightest object in the sky after the Sun and Moon, the appearance of Venus in Mesopotamian iconography as

Fig. 2.2 Detail of a stele commemorating the Babylonian King Melišhipak II (r. 1186–1172 BCE, center) presenting his daughter to the seated goddess Nanaya, who represented certain aspects of Inanna-Venus. The king holds his hand before his mouth in a respectful gesture to the goddess. Symbols evoking the gods Sin (crescent moon, top-center), Shamash (the Sun disc, upper right), and Ishtar (eight-pointed star, upper left) were included with the scene as apotropaic devices to protect the stele from desecration. Department of Oriental Antiquities, Musée du Louvre, Paris (Photo courtesy of Marie-Lan Nguyen)

an eight-pointed star (Fig. 2.2) is a kind of visual shorthand still recognizable to us today.

As civilization arose elsewhere in the region, the cultural history of Venus as a heavenly body entered a long period of stasis. Ancient Near Eastern knowledge of the night sky was transmitted to the Greeks by way of Egypt, who found something like Venus in their native goddess Isis. Herself associated with a death-and-resurrection story, Isis was also connected to calendar-keeping in one mythological explanation for the annual rise and fall of the Nile River. But the Egyptians evidently did not make the connection between the morning and evening apparitions of Venus as manifestations of the same object, calling the morning star *Tioumoutiri* and the evening star *Ouaiti*.

The Greeks adopted this view and called Venus alternately *Phosphoros* ("Bringer of Light") or *Eosphoros* ("Bringer of Dawn") and *Hesperos* ("Star of the Evening"), respectively. Much later when Saint Jerome translated the Bible into the Latin of his day, he rendered the Greek *Heosphoros* of the Septuagint and the Hebrew Helel as *lucifer* ("Light Bearer"), giving us a name for the Devil that persists to our own time.[1] Venus is the only planet specifically mentioned in the sacred writings of the early Greeks, and it appears in the works of great poets like Hesiod and Homer. During his extensive travels in Egypt, the fifth-century BCE Greek historian Herodotus identified Isis with Demeter, the Greek goddess thought responsible for the cyclical fecundity of the Earth throughout the year, indicating the continuity of the Sumerian Inanna story from the Bronze Age to classical antiquity. By the Hellenistic period (fourth- to first-century BCE), the Greeks merged the morning star/evening star aspects of the planet in the form of the existing goddess Aphrodite Ourania, although it's not clear whether that merger reflected some specific intuition about the singular nature of the two "stars."[2]

In his vast treatise on the natural world, *Naturalis Historia*, the first-century CE Roman author Pliny the Elder summarized the knowledge received from the ancient world that came before him. "Below the Sun," he wrote:

> revolves the great star called Venus, wandering with an alternate motion, and, even in its surnames, rivaling the Sun and the Moon. For when it precedes the day and rises in the morning, it receives the name of Lucifer, as if it were another Sun, hastening on the day. On the contrary, when it shines in the west, it is named Vesper, as prolonging the light, and performing the office of the moon. ... It excels all the other stars in size, and its brilliancy is so considerable, that it is the only star which produces a shadow by its rays. There has, consequently, been great interest made for its name; some have called it the star of Juno, others of Isis, and others of the Mother of the Gods. By its influence everything in the earth is generated. For, as it rises in either direction, it sprinkles everything with its genial dew, and not only matures the productions of the earth, but stimulates all living things.[3]

---

[1]See, e.g., Isaiah 14:12, "Quomodo cecidisti de cælo, Lucifer, qui mane oriebaris!" (King James Version: "How art thou fallen from heaven, O Lucifer, son of the morning!").

[2]By the "Golden Age" of Athens (fifth-century BCE), the Greeks began to demarcate two distinct aspects of Aphrodite: the celestial (*Ourania*) and the terrestrial (*Pandemos*). The former was associated with the figure later Latinized as *Urania*, the Muse of Astronomy and a daughter of Zeus by Mnemosyne.

[3]*Naturalis Historia*, 2.6. Translated by John Bostock and appearing in his edition of *Natural History* published at London in 1855.

When the Romans adopted the Greek hero Aeneas as the founder of their civilization, they incorporated a mythological origin story connecting their race to Venus: Aeneas was the son of the prince Anchises and Aphrodite. The Romans adopted the latter as *Venus*, whose name derives from a Proto-Indo-European root word meaning "to strive for, wish for, desire, love."[4]

I wanted some insight to better understand why this planet in particular was so alluring to humans who lived in times separated from our own by thousands of years, and one person immediately came to mind. Mary Stewart Adams is an expert in the folklore of the night sky who combines her extensive knowledge of ancient mythologies with the research and ideas of contemporary astronomy to critically examine the nature of the relationship between humans and the cosmos. Mary is also a passionate defender of the night from the ravages of light pollution, which separates us from witnessing firsthand the starry heavens that inspired our distant ancestors to create such beautiful stories. What, I asked her about the mythological Venus, stands out most distinctly?

"What most delights me is the consistency of attribute," Mary wrote me from her home in Michigan. "Venus seems always to be connected with love and beauty, with the ferocity of protecting what is loved, and with the feminine divine." But then she immediately turned over the cultural picture of Venus I had in mind to show the imperfections of the mythological figure represented by the planet. "It is curious to me that, although Venus was associated with devotion and reverence, her depiction in Greek and Roman mythology should include infidelity." Venus used her power to make the gods love those whom they would otherwise not. Yet by causing Venus to fall in love with Anchises, Jupiter teaches her an important lesson: she is no stronger than the other gods. Here the façade begins to crack, revealing in the divine a very human inclination toward hubris.

So what was it about the planet Venus that earlier humans found most alluring as they saw it in the night sky? Mary recalled the belief that Plato expressed in *Timaeus*, a dialog composed in the fourth century BCE in which it is suggested that each human soul literally arrived to Earth from a particular star. Because of human origins among the stars, Mary said, the ancients believed that humans had a more "inner" relationship to the planets than what we experience when we gaze upon them as though from "without." "This is really a gift given to the human being by Venus."

[4]c.f. The English word "venerate", from the Latin *venerātus*, perfect passive participle of *veneror* ("worship, reverence").

To understand why, she said, we have to look back to a time in human history during which there was no separation in the consciousness between the celestial and terrestrial worlds as we experience them now. The cosmos was widely believed to influence the destinies of individual humans, and in this astrological context the position of the planet Venus at the moment of one's birth would have been regarded as a sort of celestial statement about the capacity of that individual to rise up to the function or gift bestowed by Venus. It was a personal relationship. "Venus was considered the realm where, as the soul was incarnating, the 'decision' was made with regard to nationality and family and the type of relationship one would have in these communities, whether it be loving or estranged."

I wondered: did that relationship persist through a person's life? Mary believed so. "I imagine that Venus stirred a sense of gratitude for any type of grace that was experienced in life," she wrote, "and that its beauty and brilliance confirmed this sense that here was the spiritual divinity that could guide the human being along a path of goodness, truth, and beauty."

There is something distinctly familiar in that sentiment. Perhaps, I suggested, that is why even after the scientific study of the planet Venus began in earnest, telescopic observers reported seeing things there we don't see now, such as oceans and mountains. It was almost as if they wanted to believe that Venus as a place was much like our very familiar Earth. Was that plausible? "I think it is simply that we inherently long for affirmation that there is a mystery in our being, that not all things can be described or understood through the intellect or by logic, but that dream and imagination are sometimes better ways to address longing. In this regard it's not so much how well the idea reasons, but how the idea *feels*, and as Goethe said, 'Beauty is everywhere a welcome guest.' " To the end of antiquity, the planet Venus was seen not as a place but as the realization of an ideal, a cosmic embodiment of beauty. It took the upending of the classical world to shake that view, to be replaced by one in which Venus was an object with real, physical characteristics.

The traditional story of history books is that, after the fall of the Western Roman Empire and the decline of its influence in Western Europe, the lamp of learning was effectively extinguished as the Medieval period began and not lit again until the blooming of the Renaissance. But the light of the intellectual world didn't flicker out; rather, it simply migrated east, largely bypassing the Byzantines and establishing itself in great cities of the Islamic world like Baghdad and Cairo. In the last great era of Western astronomy before the introduction of the telescope, Islamic astronomers made remarkable observations that pushed the limits of human visual acuity. There are indications that some Islamic skywatchers with especially keen eyesight detected the crescent shape of Venus when it appeared near the Sun in the sky, giving them the hint that something was different about this wanderer that set it apart from the other planets.[5] That observation was not fully confirmed for almost 500 years, awaiting the magnifying power of the telescope.

As the Medieval period ended in Europe, the emergence of modern Western civilization unfolded mostly like we learn about in school. The still-monolithic Christian Church under the authority of the Roman Pontiff began a long, gradual decline in both its temporal and spiritual power in Europe. Europeans "rediscovered" great written works of science and literature from the ancient world, in many cases due specifically to the assiduous work of Islamic scholars who translated and transmitted them back to Europe. At the same time, explorers were setting sail on voyages of discovery that proved the world was considerably bigger than previously imagined. And with a renewed interest in learning and exploration came technological innovations that would literally transform our view of the cosmos.

Although humans discovered the basic principles of optics and were making simple lenses out of polished crystal as early as the first millennium BCE, there were few practical applications of this technology until the late Middle Ages. In the eleventh century, the Islamic scientist Ibn al-Haytham, known in the West as "Alhazen," published *Kitāb al Manāẓir* ("Book of Optics"). This seven-volume treatise summarized essentially everything that Alhazen could find about knowledge of optics in the ancient world, relying particularly on *Optics*, a work by the second-century CE Greco-Egyptian astronomer Claudius Ptolemy, and an account of the anatomy of the human eye by Ptolemy's contemporary, the Roman physician Galen. Alhazen's *Book of Optics* was rendered into Latin as *De Aspectibus* ("On the Aspects") by an unknown translator toward the start of the thirteenth century, and it quickly became influential in Europe.

European opticians, influenced by Alhazen, were experimenting with grinding and polishing glass lenses by 1250, adapting them to applications like vision-correcting spectacles soon after. The magnifying capabilities of lenses were exploited to produce the first monoculars, or spyglasses, roughly 300 years later. And while we don't know who invented what we now call the telescope, history records that the first patent application for one was submitted by

[5]There is some debate over whether people with unusually good visual acuity may have been able to resolve the crescent shape of Venus during certain of its phases. q.v. Henry MacEwen, 1895, The Visibility of the Crescent of Venus, *Journal of the British Astronomical Association*, 6, 34–35; and W.W. Campbell, 1916, Is The Crescent Form of Venus Visible to the Naked Eye?, *Publications of the Astronomical Society of the Pacific*, 28(2), 85.

Hans Lippershey (1570–1619), a spectacle maker from Middelburg, the capital of the Dutch province of Zeeland. In October 1608, Lippershey applied for a patent on an instrument whose purpose was "for seeing things far away as if they were nearby." It's unclear exactly what inspired him, but one account has it that Lippershey stumbled upon the idea for a spyglass after watching children in his shop holding up pairs of lenses that magnified the image of a weather vane atop a distant roof.

The entrepreneur in Lippershey smelled a government contracting opportunity: with this device, for example, one could gather important intelligence about invading navies while they were still far out at sea. While Lippershey's patent application was ultimately unsuccessful, the idea of the spyglass caught fire and rapidly spread across Europe. It was only a matter of time before someone turned a telescope skyward.

Galileo Discovers the Phases of Venus

In early 1609, the Florentine scientist Galileo Galilei (1564–1642) was traveling in Venice when he first heard of something then called the "Dutch perspective glass." Despite having never examined one in person, immediately upon his return home to Padua, he set about fashioning his own example. He toyed with the design, fitting different lenses into a leaden tube until he achieved an acceptable result. Sensing some potentially sensational attention for the benefit of his academic career, Galileo hurried back to Venice, telescope in tow, and presented the device to the Venetian Senate in the presence of Doge Leonardo Donato. On August 21, from an elevated viewing area in the bell tower of St. Mark's Cathedral, Galileo publicly demonstrated the telescope before the Doge and members of the Senate (Fig. 2.3), who in turn gave him a permanent position at the University of Padua and doubled his salary.

Galileo continued to refine the design of his telescope, improving its performance through the autumn of 1609. In the waning weeks of that year into the first frosty nights of 1610, he pointed it toward the Moon and Jupiter. The latter was readily observable, high in the evening sky in the constellation Taurus throughout the winter of 1609–1610, and he quickly found (and was the first to describe) its set of four large moons. They appeared to Galileo as featureless points of light slowly wandering back and forth along an imaginary line passing through the planet itself. Since the objects were bound to follow Jupiter wherever it wandered across the sky, Galileo quickly concluded that they must be in orbit about the planet and that our point of view was more or less in the same plane as their orbits.

Fig. 2.3 *Galileo presents the telescope to the Senate of Venice* (c. 1840) by Luigi Sabatelli (1772–1850), housed at the Tribune of Galileo in Florence, Italy. Photo by Wikimedia Commons user Sailko, licensed under CC BY-SA 3.0

The existence of the "Galilean Moons" was an unexpected repudiation of the cosmology of Aristotle (384–322 BCE), who held that the Earth was the fixed center of the universe. What was Jupiter doing with what appeared to be its own miniature planetary system? The delicate dance of the little spots about Jupiter from night to night was strongly reminiscent of the Sun-centered model of Nicolaus Copernicus (1473–1543) if one simply exchanged Jupiter for the Sun, upending the view in Galileo's time that Jupiter was fully subordinate to the Sun in the hierarchy of the Solar System.

Galileo, very aware of the source of his livelihood, immediately named the curious objects after his patron, Cosimo II de Medici (1590–1621), Grand Duke of Tuscany. In the dedication of his groundbreaking treatise *Sidereus Nuncius*, published a month later, Galileo entreated the Grand Duke to "behold, therefore, four stars reserved for your illustrious name … that make their journeys and orbits with a marvelous speed around the star of Jupiter … like children of the same family." The implied comparison of himself to the king of the Roman gods probably flattered the Grand Duke.

Jupiter was the only bright planet observable in the evening during the first half of 1610; Mercury and Venus were gradually lost in the glare of morning

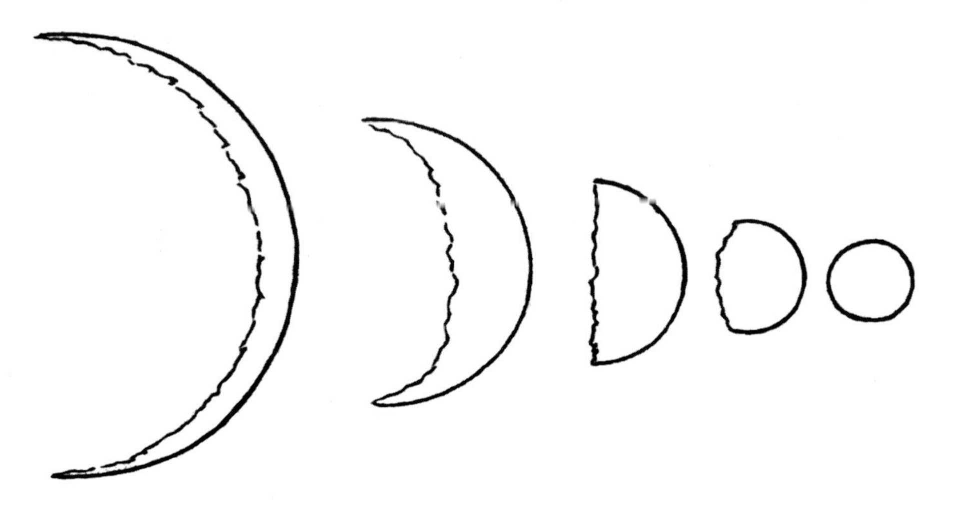

Fig. 2.4 Galileo's drawings of his observations of the phases of Venus, published in *Il Saggiatore* (The Assayer), 1623

twilight around the time Galileo was making his Jupiter discoveries. But Venus reappeared in the early evening during late summer, and near the beginning of October Galileo found it with his telescope. He saw no indication of moons orbiting Venus as they did Jupiter. The planet appeared unremarkable to him, a dazzling, featureless circle that might as well have been a perfect mirror. Yet his intuition led him to keep watching.

As the weeks passed he saw the shape of Venus gradually begin to enlarge, distort, and elongate as the planet approached Earth in its orbit (Fig. 2.4). By December, the conclusion was inescapable: even at the poor resolution of Galileo's simple telescope, Venus could no longer be fairly described as a circular disc. Just as in the case of the little Solar System he found centered on Jupiter, the dominant cosmology of his time made no accommodation for the observed fact that Venus exhibited phases like those of Earth's Moon.

Given how quickly others in Europe were fashioning telescopes for astronomical use, Galileo well knew that his monumental discovery about Venus could easily be scooped. While he realized that he wouldn't have the luxury of observing a full cycle of the phases before publishing the details, he wanted to make a select few astronomers aware of what he had seen. To buy some time, Galileo followed a common practice of the era: he encrypted his result into a Latin anagram. On December 11, he wrote to the German astronomer Johannes Kepler (1571–1630) "Haec immatura a me jam frustra leguntur o y," which translates roughly as "This was already tried by me in vain too early." However, the extra "o" and "y" at the end of the statement were a clue to

Kepler that Galileo was transmitting a secret message to him. It turned out that Galileo wasn't all that good at constructing anagrams, and couldn't find a place to insert the extra letters for his real message to the recipient. Kepler took the hint and unscrambled the letters to form a new sentence: "Cynthiae figurae aemulatur mater amorum" ("The mother of lovers imitates the shapes of Cynthia"). Knowing that the "mother of lovers" was Venus, Kepler quickly understood: Venus undergoes a regular cycle of phases like the Moon. It was big news indeed.

On December 30, Galileo wrote to the German Jesuit astronomer Christopher Clavius (1538–1612). Feeling increasingly confident that the apparent phases of Venus were in fact real, he described his observations in plain language for the first time. "When Venus began to be visible in the evening sky," Galileo recounted:

> I started observing it and saw that its figure was circular, though extremely small. Afterwards, I saw [Venus] growing in magnitude significantly, though always maintaining its circular shape. Approaching maximum elongation, Venus began to lose its circular shape on the other side from the Sun and within a few days had acquired a semicircular shape. This shape it maintained for a number of days. More precisely, it maintained [this shape] until it began to move toward the Sun, slowly abandoning the tangent. It now begins to assume a notable corniculate shape. Thus, it will continue to decrease during the period in which it remains visible in the evening sky.[6]

Galileo remained puzzled by this behavior. A crescent shape was in no way reconcilable with the reigning view of Ptolemy, whose cosmology aligned clearly with Aristotelian principles. In the Ptolemaic system of Galileo's time, Venus was assumed to be "above" the Sun in a hierarchy of objects whose perfectly circular orbits centered on the Earth. But the observed motions of the planets didn't comport with the geocentric ideal, so Ptolemy bent the model until it fit the data. He resorted to a series of "epicycles," small, circular orbits-within-orbits executed about the line of the larger Earth-centered orbit. None of this had any basis in the fledgling notion of physics as it existed in antiquity, and a 1,000 years would pass before keen observers noticed that the simplest explanations for natural phenomena tend to be the right ones. Starting with a conclusion, Ptolemy added more and more moving parts until it all seemed to work well enough.

[6]From *Le opere di Galileo Galilei* in 20 volumes, Edizione Nazionale, ed. Antonio Favaro, Florence, 1890–1909. It is quoted in translation by Palmieri (2001).

Galileo knew that a full or gibbous Venus was consistent with an orbit "above" the Sun in the Ptolemaic hierarchy, while a crescent Venus was only compatible with an orbit in which the planet was forever "below" the Sun. The trouble, of course, was that Venus exhibited *both* crescent and gibbous phases, which could not be reproduced within the framework of a geocentric Solar System. Ten more years of thought would pass before Galileo finally convinced himself that Venus orbited not the Earth but rather the Sun itself, and therefore the Ptolemaic model was fundamentally wrong. He presented the conclusion in *Il Saggiatore* ("The Assayer"), published at Rome in October 1623, one of the first written works on what we would now call the scientific method.

Telescopic observers of Venus in the early seventeenth century were tremendously limited by the various imperfections of early instruments, including the low angular resolution of their small objective lenses and the poor quality of the glass used to make them. These telescopes suffered from chromatic aberration, an optical effect resulting from the tendency of glass to bend the paths of light rays through different angles according to their colors. This meant that the red light rays came to one particular focus behind a lens while the blue rays came to a different focus, leading to brilliantly colored haloes around astronomical objects. The effect was exaggerated for bright objects like the planets, making it difficult to distinguish subtle differences in color and intensity. In many respects, the resulting images weren't just aesthetically unpleasant: they could actually induce the observation of apparent phenomena that didn't really exist.

The early adopters of "Galilean" refracting telescopes understood that the colors were ephemeral and produced in the optics themselves while they struggled to find innovative ways to correct for the effect. Around 1730, an English lawyer and amateur inventor named Chester Moore Hall tinkered with lenses, finding that a combination of dissimilar lens materials and optical surfaces of different kinds of curves could overcome chromatic aberration. But even after improved lens designs began to yield planetary images with better fidelity, the shimmering disc of Venus gave up no new secrets. No matter how impressive the image, the planet remained utterly featureless. Was it a perfect sphere, devoid of topography? Was it covered in water? Was its apparent surface actually a uniform layer of clouds? Neither bigger telescopes nor higher magnification yielded any additional detail that might help answer any of these questions. For decades, Venus gave up no more of her secrets.

And what of the Earth's Moon, in comparison with Venus? Unaided human eyes turned to our Moon see a lot, given that the angular diameter of the full Moon is some 30 times that of Venus when it approaches Earth most closely. At its widest, the human eye pupil opens to a diameter of about 8 mm when fully sensitized to very dim light. A lower limit to the eye's power to resolve objects of small angular size is set by the so-called diffraction limit, a property imposed by the wave nature of light. A keen observer under ideal conditions, working at the diffraction limit of the human eye, might resolve features with angular diameters as small as 15 to 20 seconds of arc, each second being 1 part in 3600 of 1 angular degree. At the minimum possible distance from the Earth to the Moon, roughly 363,000 km, an ideal human observer might be able to discern features corresponding to linear sizes as small as 30 km.

But the situation in practice is complicated by glare and contrast effects, which tend to degrade visibility and therefore render small objects more difficult to see, as well as the optical distortion caused by Earth's often turbulent atmosphere. It's also the case that the Moon is sufficiently bright, even against a dark background sky, that the pupil of the eye constricts in trying to handle the large dynamic range of the illuminated surfaces in view, so the angular resolving power set by diffraction is decreased. However, some lunar features such as its dark lava plains, or "maria," are easily visible by virtue of their contrast against lighter-colored surroundings. These tend to be hundreds to thousands of kilometers across.

In any case, it's fair to say that a visual observer with typical eyesight might be able to make out something on the Moon with a linear size of about 300 km, and one with especially keen eyesight might be able to do 100 km better than that. I've tried it myself many times; for example, I can readily make out the roundish Mare Humorum (diameter 390 km) and Mare Nectaris (340 km) as clearly separate from their surroundings, whereas I have never confidently detected the crater Grimaldi (175 km) despite its dark, lava-flooded floor and the bright, adjacent highlands. These features are indicated with labels on an image of the full Moon in Fig. 2.5. The much higher angular resolution of telescopes, rather than their light-gathering capacity, is what lends so much more detail to views of the Moon than our eyes could ever reveal.[7]

Applying the same logic to Venus, given its apparent angular size and minimum distance (about 38 million kilometers), a modest telescope of a few tens of centimeters' aperture could reveal features on Venus as small as about 100 km across. Since that figure is comparable to the size scales of various terrestrial landforms, and if Venus were a world not unlike Earth, we would naively expect to see through our telescopes a mottled surface with

[7] A good overview of the resolving power of the unaided eye as applied to lunar features can be found in Joseph Ashbrook's article "Lunar Studies Before the Invention of the Telescope," originally published in *Sky and Telescope*, June 1962, p. 322. It was republished in the compilation *The Astronomical Scrapbook: Skywatchers, Pioneers and Seekers in Astronomy* (Cambridge, Mass.: Sky Publishing, 1984, p. 233).

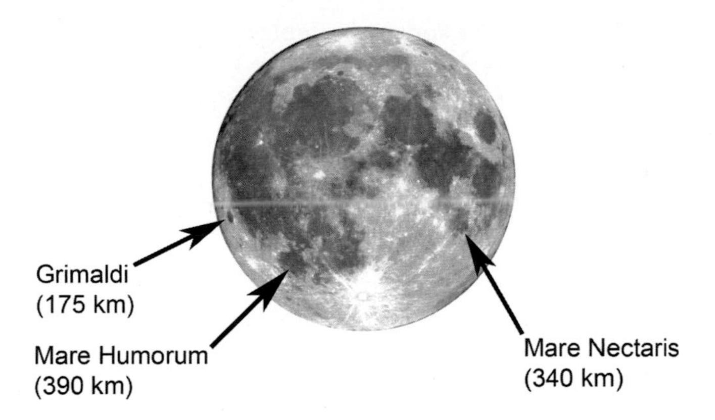

Fig. 2.5 Lunar features near the spatial resolution threshold of the unaided human eye at the distance of the Moon referred to in the main text. Photo by the author

distinct continents, seas, mountain ranges, and even polar ice caps. Yet the telescope revealed none of these things to Galileo and his contemporaries, even though later as telescope quality improved observers began to report shadowy markings most consistent with cloud patterns. That is, until around the middle of the seventeenth century.

In 1651 another Italian, Giovanni Battista Riccioli (1598–1671), wrote about telescopic observations of Venus he made several years earlier: "As chance would have it, … a partial circle of light could be seen enclosing the otherwise dark side of the disc, brightening it particularly toward the eastern limb." It seems at first glance like a straightforward and otherwise uninteresting account of what Riccioli saw on the night of January 9, 1643, through a simple refracting telescope that his friend, the lawyer Francesco Fontana (*c*. 1580–1656), loaned him. In fact, Riccioli saw something entirely without precedent in the history of astronomy up to that point in time.

It was only a generation after the invention of the telescope for astronomical purposes, prior to which nothing was known about the planets other than they constantly changed their apparent positions on the night sky unlike the seemingly fixed stars beyond. Until Galileo turned his simple spyglass on the night sky for the first time, no one had any good idea what the planets were like, much less that they were worlds unto themselves. The moons of Jupiter and the phases of Venus and Mercury confounded European observers in their attempts to understand the observations in light of the dominant cosmology of the era. But, just as in Giovanni Schiaparelli's time, telescopes showed what they did, and they left certain impressions on the people who viewed the planets through them.

The optical shortcomings of Fontana's simple refracting telescope were clear to Riccioli on the night he first saw the disc of Venus in its entirety; the planet "was, compared to the Sun, red to yellowish and, on the side facing the direction of the Sun, a blue-green: but those colors were most likely caused by the particular variety of glass used in the telescope optics." And yet there was the otherwise unilluminated "night" side of Venus, glowing unmistakably against the darker night sky surrounding it.

The same sensation was neatly summarized by the German astronomer Christfried Kirch (1694–1740) as he peered through the eyepiece of his 26-foot telescope at the Berlin Observatory on the evening of Friday, March 8, 1726: *Ich konnte das tunkle Theil Veneris erkennen* ("I could see the dark part of Venus").[8]

Riccioli was the first to describe what later became known as the Ashen Light (Fig. 2.6), a ghostly emanation from the nighttime hemisphere of Venus, a complete explanation for which has eluded science for almost 400 years.[9] Writing toward the end of the nineteenth century, the Irish linguist and journalist Ellen Mary Clerke (1840–1906) gave perhaps the most poetic description of the Ashen Light to date. It is the circumstance, Clerke wrote:

> when the shadowy form of the full orb is seen to shine dimly within her crescent. . . . More wonderful still, this "glimmering sphere" is then crowned, as with a silver halo, by a thin luminous arch, forming a secondary sickle facing the one nearest the sun, and doubtless due to the refraction of his rays round the globe of the planet, through the upper regions of her twilit atmosphere.[10]

We don't know for sure what it is or even *if* it is. The Ashen Light has become a sort of "Loch Ness monster" of planetary astronomy reported only by eyewitnesses. It has firmly (and, perhaps, conveniently) refused to submit to convincing photography or digital imaging in the time since those processes were invented. And its visibility seems in no way predictable, working strongly against efforts to independently establish its very existence. "There is one characteristic of the phenomenon abundantly verified by the numerous observers who have recorded it," the editors of the scientific journal *Nature* noted in their "Astronomical Column" of June 1876, "which cannot be

[8]Quoted by Eduard Schönfeld in *Astronomische Nachrichten*, Vol. 67, No. 1586, p. 27, 1866.

[9]The same phenomenon is sometimes referred to as the "secondary light" of Venus, especially in scientific journals of the nineteenth century, to distinguish it from the "primary light" reflected directly from the Sun.

[10]1893, *The Planet Venus*, London: Whiterby and Company, 15–19.

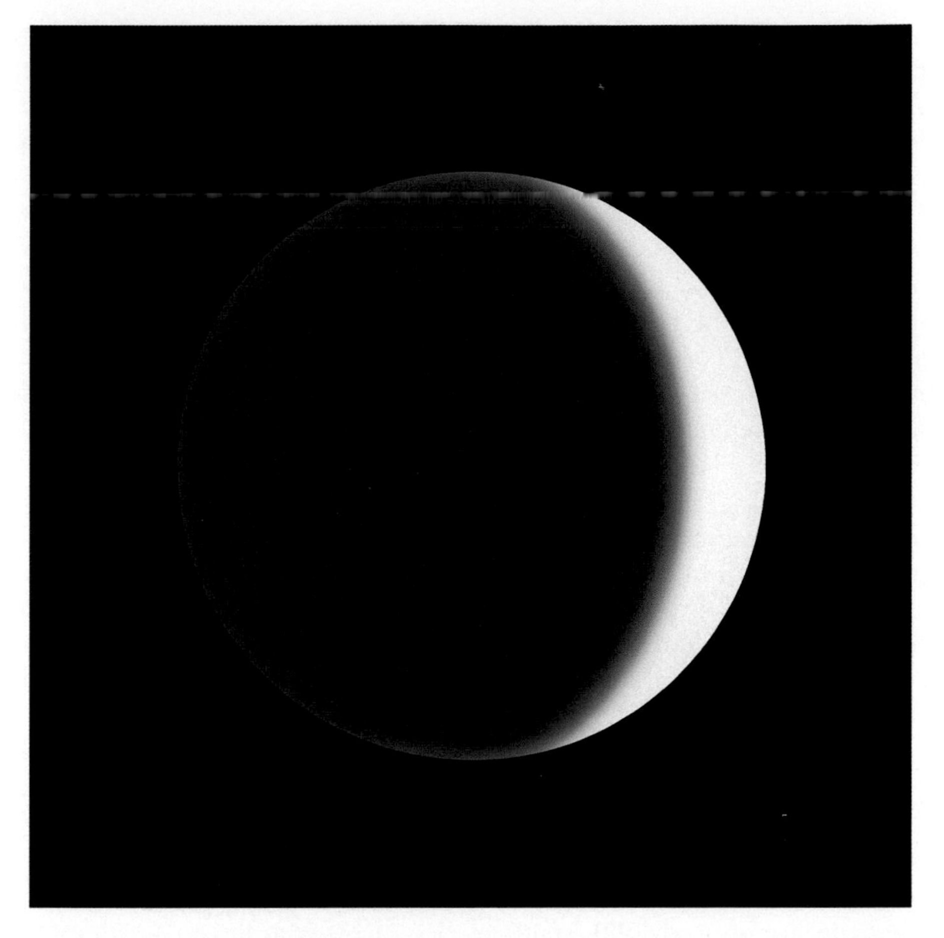

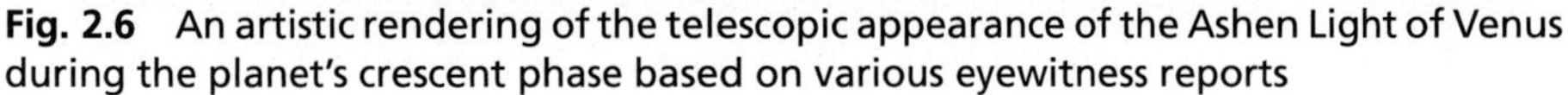

Fig. 2.6 An artistic rendering of the telescopic appearance of the Ashen Light of Venus during the planet's crescent phase based on various eyewitness reports

overlooked in our endeavours to arrive at its true cause, viz., its intermittent or only occasional visibility." The Ashen Light is at the same time seductive and frustrating, defying all attempts at definitive elucidation while simultaneously resisting assignment to the scrap heap of astronomical myth. It seems to forever exist in the shadowy twilight between truth and fiction.

Yet there have been periods of time in astronomical history in which "the Ashen Light has been glimpsed by almost every serious observer of Venus," wrote the great English amateur astronomer, Sir Patrick Moore (1923–2012), in 1982. Moore himself was a prolific observer of the Ashen Light, recording the phenomenon on dozens of occasions from the 1950s nearly until his death. Convinced of its reality by sensing the Light repeatedly with his own eyes, Moore argued that the phenomenon positively cried out for investigation: "Genuine or not, the Ashen Light has been seen so frequently, and by so many experienced observers, that it has to be explained in some fashion. ... [T]he observations of the Light are too numerous, and too concordant, to be dismissed."

The Revered Thomas William Webb, whose *Celestial Objects for Common Telescopes* (1859) remained essential reading for the British amateur astronomer for over a century after its publication, described the Ashen Light as a "truly unaccountable appearance" of Venus that was, nevertheless, "remarkably well attested."[11] Even Percival Lowell, who as far as we know did not witness the Ashen Light himself—or, at least, didn't record it in his voluminous observing notebooks—commented in 1909 on both its persistence in observational history and the failure of all attempts to find a sufficiently scientific cause: "The phenomenon has seemed the weirder for the difficulty of explaining it." *Sky and Telescope* magazine's Thomas Dobbins wrote in 2012 of the Ashen Light's outright rejection by many modern astronomers. "Many authorities dismiss the Ashen Light as an optical illusion," Dobbins noted. "Skeptics find it hard to believe that the phenomenon has not been recorded in images or spectrograms obtained with large Earth-based telescopes, let alone by the various spacecraft that have orbited Venus."

How, then, would one go about finding a solution to such an ephemeral problem? One place to start is the historical record and the words of Ashen Light eyewitnesses who committed their recollections to paper. Certain patterns emerge by surveying accounts of their observations. For instance, reported sightings are not evenly distributed through time. Figure 2.7 shows the frequency of nearly 500 reports by decade between the early eighteenth and the early twenty-first centuries, presented in such a way that the number in each bin is a percentage of the total number of reports. One-quarter of all recorded sightings happened in the 1950s, which may bear significantly on physical interpretations of the phenomenon. But it may simply be that more observers were looking at Venus during that particular decade than others and that the spike in the distribution hints at observer bias. This, too, could explain the number of reports from the second half of the nineteenth century, as planetary astronomy was rather in vogue during the late Victorian era. And relatively few telescopes existed in the world in the period of time represented by the graph's earliest decades, during which very few people anywhere were looking carefully at Venus.

[11]1873, 3rd ed., London: Longmans, Green & Co., 55.

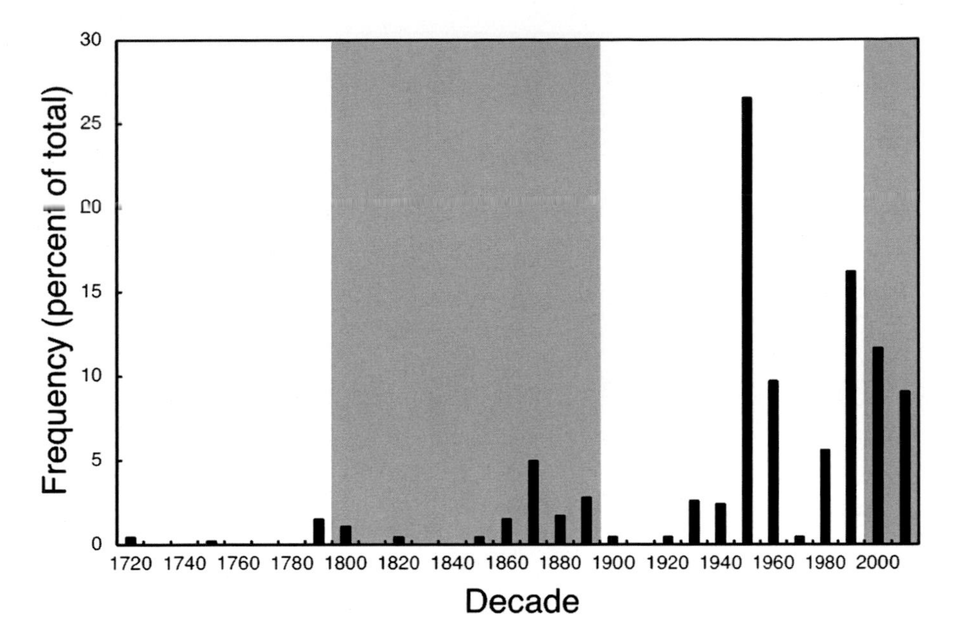

Fig. 2.7 A histogram showing the frequency of 465 Ashen Light observations between 1721 and 2017. The columns are organized by decade and labeled according to the first year of each. The number of observations in each decade is divided by the total and the result converted into a percentage. The gray-shaded regions set the nineteenth and twenty-first centuries apart from the eighteenth and twentieth centuries for clarity

One might ask, for example, whether there is any connection with the magnetic activity cycle of the Sun. If the Ashen Light is caused by anything like the Earth's aurorae, then its appearances might correlate with periods of time in which the Sun is more alive with flares and mass ejections that spew charged particles outward into the space surrounding the planets. If a correlation between the intensity of solar activity and the likelihood of Ashen Light observations were found, it would strengthen a Sun-Venus explanation.

To help search for any such connection, Fig. 2.8 adds average sunspot counts by decade to the distribution of Ashen Light sightings from Fig. 2.7. The data are inconclusive on this point; still, there are hints in this figure that something is indeed up. The apparent spike in the number of Ashen Light sightings in the 1950s correlates with one of the strongest solar maxima of the past 250 years. Given that this peak also coincides in time with the International Geophysical Year of 1957–1958 (Chap. 6), it may be indeed that more people than usual were observing the planets during those years and therefore more likely to note anything unusual about the telescopic appearance of Venus. Note further that the declining trend of sightings since the 1990s mirrors the overall decline in

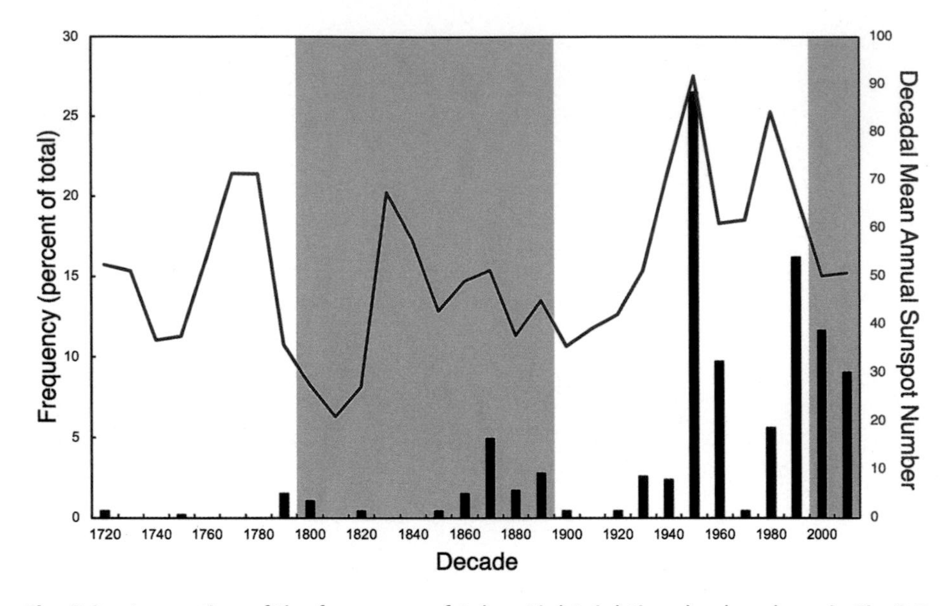

Fig. 2.8 A recasting of the frequency of Ashen Light sightings by decade, as in Fig. 2.7, but with the addition of the average annual number of sunspots by decade shown in red on the secondary vertical axis. For visual convenience, gray-shaded regions again separate the centuries

sunspot counts during the past three decades. Any simple connection between solar activity and the likelihood of the Ashen Light being seen is therefore neither ruled in nor out by the data.

Attempts to connect appearances of the Ashen Light with various influences are frustrated by its fundamental unpredictability. "Outbreaks," during which many observers widely scattered in geography report seeing the Ashen Light night after night for weeks on end, punctuate periods in which one finds no reports in the literature for as long as a decade. Richard McKim, director of the British Astronomical Society sections on Venus, Mercury, and Mars, noted that "sometimes the Ashen Light is visible for many days in succession, as in 1940, 1953, 1956, and 1957. At other times, despite excellent observing conditions, it remains elusive." The existence of outbreaks just raises more questions: was it the case that whatever causes the Ashen Light wasn't active during that time? Or could it just be that no one was looking?

It also doesn't seem to be a simple trick of the telescope, which for any particular instrument and observer would manifest itself every time Venus were observed. It would then be possible to stage conditions under which Ashen

Light is reported by any user of a given telescope.[12] That no one has fully reproduced the effect, much less captured a photograph of it, lends credence to the notion that the Ashen Light is at least not a purely instrumental artifact.

That it might have something to do with the atmosphere of Venus is supported by observations of a so-called 'limb brightening' effect in which the intensity of the Ashen Light seems to tick upward somewhat just before the reaching the edge of the planet's disc, resulting in the appearance of the "silver crown" described by Ellen Mary Clerke. Imagine a spherical planet whose solid surface is covered by a luminous atmosphere of uniform thickness, making the atmosphere a kind of concentric "shell" around it. If the atmosphere emits light evenly throughout its bulk, then its brightness in any given direction depends on how much of the atmosphere one looks through. If the observer is hovering over the planet and looking directly down toward its center, the light-emitting part is "one atmosphere" in thickness and yields a certain brightness. But as the observer's gaze moves from the center of the planet's disc to the edge, the effective path length through the atmosphere increases; at the planet's limb, one sees through many "atmospheres" of thickness, which appears consequently brighter. Since the path length is longer at the limb than toward disc center, a greater amount of atmosphere contributes light in the direction of the observer, who sees the apparent brightness of the planet increases from center to limb.

The German astronomer Hermann Joseph Klein (1844–1914) described this effect in the context of Ashen Light observations:

> If a planet enveloped by an atmosphere is illuminated from the outside, [and] if the envelope of the vapor does not absorb the light like a dense layer of fog, the edge of the planetary disc must appear a little brighter than the center. This is indeed the case with the secondary light of Venus, when the observers of the phenomenon saw the edge of the planetary disc stand out sharply from the heavens.[13]

---

[12] See the experiments exactly to this point reported by Sheehan, Brasch, Cruikshank, and Baum in the *Journal of the British Astronomical Association*, Vol. 124, No. 4, pp. 213–214 (2014). In dismissing the validity of existing photographs claiming to have recorded the Ashen Light, the authors note that "none of the putative images of the actual AL [Ashen Light] can be regarded as convincing. They appear to be due to filter leakage, crescent glare, and excessive image processing used to bring out dark-side detail. It is, of course, difficult to prove a negative."

[13] 1873, *Handbuch der allgemeinen Himmelsbeschreibung: Vom Standpunkte der kosmischen Weltanschauung*, Braunschweig: Friedrich Vieweg & Son, 69.

The English surgeon, antiquarian, and amateur astronomer Charles Leeson Prince (1821–1899) wrote of seeing a "phosphorescent flitting of light around the edge of the entire disk" of Venus from his home in Uckfield, Sussex, on both the 25th and 30th of September 1863.[14] Sir William Herschel (1738–1822), the famed Anglo-German discoverer of the planet Uranus, made a remarkable series of observations of the Ashen Light at Observatory House in Slough, Berkshire, in April 1793. "The light of Venus is brighter all around the limb, than on that part which divided the enlightened, from the unenlightened part of the disk," he wrote in his notes for the evening of April 9.[15] Later in the same night, the planet's limb showed "a luminous margin …like a bright bead, of nearly an equal breadth all around." On the night of the 20th, Herschel saw "a narrow luminous border all around the limb, and the light afterwards diminishes pretty suddenly." During an April 1876 meeting of the Royal Astronomical Society, William Noble (1828–1904) recounted seeing on an unspecified date "some years back …unmistakably the whole body of Venus, with the illuminated crescent like a hair of light around it."[16] Each such account clearly suggests limb brightening at work.

So it would seem at first glance that the mystery is solved: Venus must be surrounded by some kind of translucent material that is faintly self-luminous, explaining both the overall phenomenon of the Ashen Light and the observed limb brightening. And while Venus is very clearly possessed of a substantial atmosphere, it has been shown to emit no considerable quantity of light to which human eyes are especially sensitive. A uniformly luminous medium enveloping Venus therefore can't be the right answer.

The picture is further clouded by the existence of not one but *two* broad manifestations of the Ashen Light in the historical record, one seen in daylight and the other against a distinctly darker night sky. In the former case, observers describe some variety of discoloration of the planet's disc, making it stand out against the sky background by way of color contrast, or the appearance of the portion of the disc not directly illuminated by the Sun in such a shade of deep blue that it seems darker than the sky (Fig. 2.9). These instances have been labeled "negative Ashen Light," to differentiate them from their "positive" kin

[14] 1863, Observations of Venus at the Inferior Conjunction, *Monthly Notices of the Royal Astronomical Society*, 24, 25.

[15] 1793, Observations on the planet Venus, *Philosophical Transactions of the Royal Society of London*, 83, 209–213.

[16] 1876, *The Astronomical Register*, 14, 110. Noble stated clearly that he *didn't* note this effect during any of his previous observations of Venus. On September 26, 1870, for example, he "quite failed to see the dark body of the planet, which under analogous conditions has always been visible enough before in a constricted field." (1871, *Monthly Notices of the Royal Astronomical Society*, 32, 17).

in which the night side of the planet seems brighter than the surrounding sky and is typically only seen when the sky is astronomically dark (or nearly so).

Patrick Moore was clear about his distaste for applying "Ashen Light" to any daytime observation, which he labeled "obviously a contrast effect." Although Moore noted that there clearly exist observations of more than one type, "the term 'Ashen Light' should properly be restricted to the faint luminosity of the night part of the disk, and should not be extended to cover reports of the night side being seen as darker than the background."[17] Moore held that the nighttime version of the effect was likely a true glow of some kind in the atmosphere of Venus, while the daytime version could be explained away as a trick of light and optics.

What about an explanation that is a little closer to home, in a manner of speaking? An obscure German astronomer evidently known only to history as "F. Schorr" published a treatise in 1875 called *Der Venusmond: Und Die Untersuchungen Uber Die Fruheren Beobachtungen Dieses Mondes* ("The Moon of Venus: And the Investigations on the Earlier Observations of This Moon") in which he argued strongly that Venus must have an undiscovered natural satellite that accounts for some occasional oddities of its appearance. Although this putative moon was no more real than the Martian *canali*, its existence might have explained the Ashen Light in a very straightforward way: just as the Earth reflects sunlight to the night side of the Moon, which is then reflected back to the Earth and seen as a similar "ashen" glow, an adequately large natural satellite on the night side of Venus might reflect sunlight to the cloud tops of Venus, which then reflect the same light back in the direction of Earth.

To support his argument, Schorr undertook an analysis of as many published accounts of the Ashen Light as he could find. "The perception of the dark side of Venus," he wrote, "can be classified, according to observations, into those in which the whole night-side was visible, and in others in which only a part of it was visible in the light peculiar to this phenomenon." His four categories were:

(i) The entire night side is visible;
(ii) The entire night side is visible during a (total) solar eclipse;
(iii) Partial visibility of the night side; and
(iv) Unequal brightness of the two horns of the crescent Venus.

[17] 1962, *Journal of the British Astronomical Association*, 72(6), 265.

Fig. 2.9 An artistic impression of the "negative" form of Ashen Light, seen exclusively against the daytime sky, in which the unlit portion of the Venus disc appears darker than the sky background

Schorr also looked for some connection between appearances of the Ashen Light and other physical circumstances of Venus and the Earth on the days and times of the observations. Of the few dozen reports he analyzed, he concluded that they "prove sufficiently that the perception of the secondary light is not bound to any time of day or evening; solar eclipses have no influence on them, only they are dependent on the time of inferior conjunction." The latter term refers to the circumstance in which Venus makes its closest approach to Earth during one orbit around the Sun (Fig. 2.10), appearing so close to the Sun in terrestrial skies that visual observers must take extreme caution to shield their

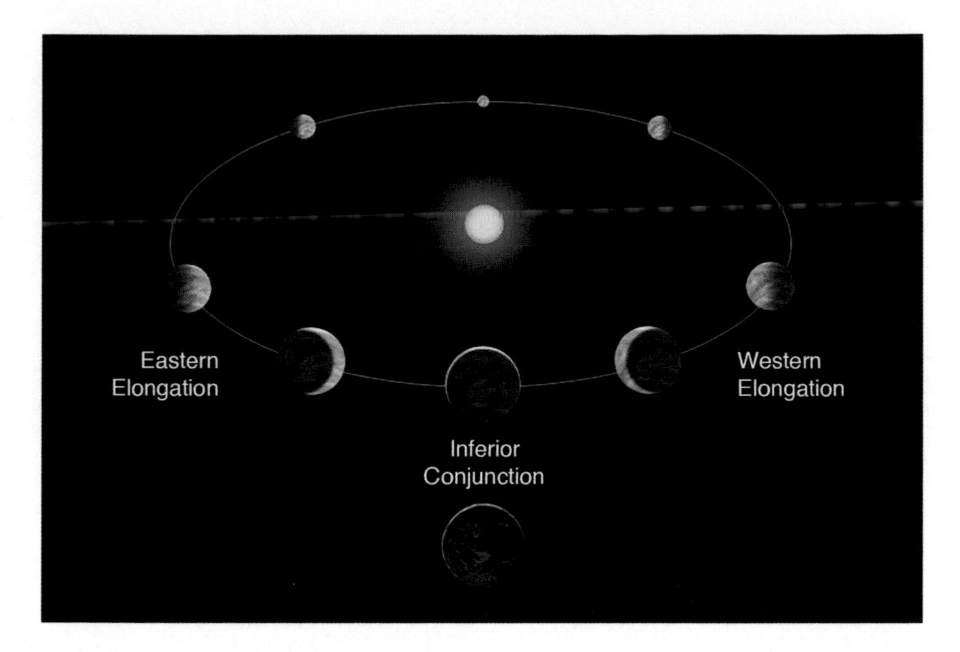

Fig. 2.10 Venus goes through a series of phases, as seen from Earth (bottom), which depend in appearance on where the two planets are situated with respect to one another. This diagram is labeled in order to illustrate the meanings of the terms "eastern elongation," "western elongation," and "inferior conjunction." Adapted from the original work of Wikimedia Commons user Vzb83 and licensed under CC-BY-SA-3.0

telescopes from the direct light falling into their open apertures in order to avoid eye damage.

At inferior conjunction Venus is nearest the Earth in its current orbital cycle, so it appears to us at its greatest angular size. At the same time, the lit crescent shrinks to its thinnest, and Earth is presented with the greatest view of the planet's night side. From his own observations of the Ashen Light in early 1806, the German astronomer Johann Hieronymus Schröter (1745–1816) concluded that the phenomenon could *only* be seen near inferior conjunction, arguing that a typical combination of observational circumstances prevented its appearance otherwise:

> [T]he only possible observations [of this phenomenon] are limited to relatively few days around the crescent phase of Venus under otherwise favorable conditions, and that even for the most diligent observer this is therefore a coincidence, because twilight and the very dim light of the night side of Venus do not permit such an appearance. It is therefore limited only to certain elongation angles of the planet.

It is true that the influence of reflected light from a moon illuminating the Cytherean cloud tops would be greatest when Venus and Earth approached each other most closely. And while at face value the "planetshine" hypothesis seems promising, all searches have failed to turn up evidence for a large moon. It's now accepted that any natural satellites of Venus must be so small as to have escaped detection altogether, one consequence of which is that there exists no natural sunlight reflector of sufficient size that can account for any faint glow emanating from the night side of the planet.

But what about the original notion underpinning Schorr's idea: are Ashen Light sightings more common near inferior conjunction than at other times? In other words, is an observer more likely to see it as Venus slips past the Sun than in other instances during which the planet is in some other part of its orbit? It turns out the Ashen Light *is* significantly more likely to be seen in the few weeks ahead of inferior conjunction than shortly afterward. Figure 2.11 illustrates this effect. It presents the same set of observations sorted by decade in

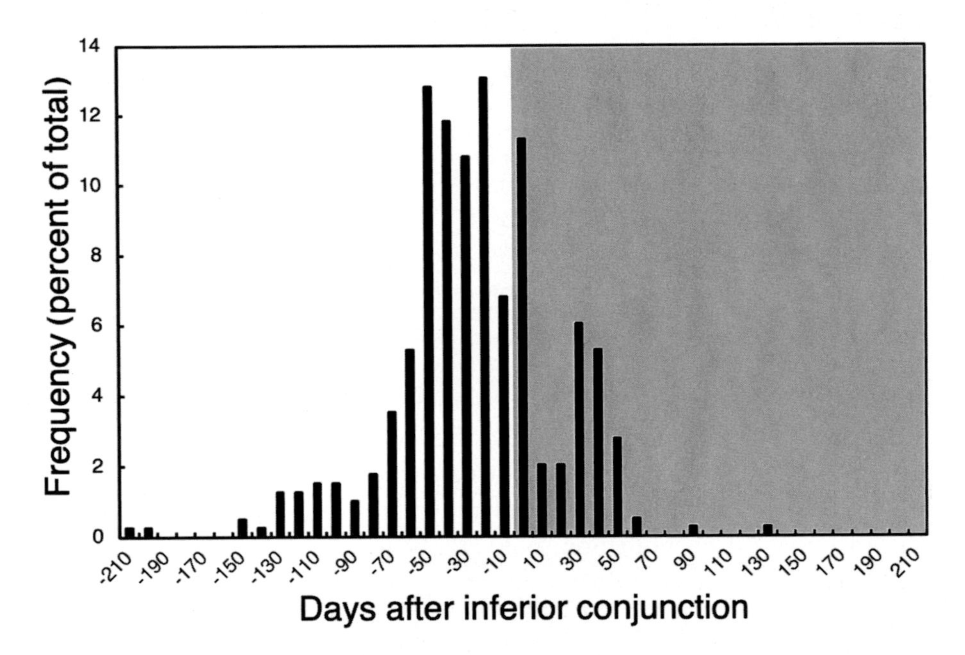

Fig. 2.11 A histogram, using the same database of observations used to make Figs. 2.7 and 2.8, showing the frequency of reports of the Ashen Light with respect to the number of days before or after the inferior conjunction at which it was observed. The gray-shaded area separates observations after inferior conjunction from those before. This naturally divides the histogram into two regions: one in which Venus is most readily visible in the evening (unshaded) from another in which observations are most conveniently made during the hours before sunrise (shaded)

Fig. 2.7, but this time the histogram bars represent the number of observations of the Ashen Light in the weeks before and after inferior conjunction. And the distribution is distinctly lopsided.

Far from conjunction, when the angular size of the disc of Venus is small, its phase relatively large and its location close in the sky to the Sun, there are few historical records of the Ashen Light; it's just not an easy observation to make under the circumstances. As Venus approaches Earth, its disc becomes larger and the phase smaller, showing progressively more of the side of the planet not directly illuminated by the Sun. It also "elongates" east or west of the Sun in the sky, depending whether it is approaching the Earth or receding from it around the time of inferior conjunction, making it generally easier to observe.

The figure shows that Ashen Light sightings peak about a month before inferior conjunction and drop off quickly thereafter; a secondary peak follows about 1 month later as the planet heads for its greatest elongation west of the Sun. Whatever controls the distribution of sightings with respect to the date of inferior conjunction can't be the apparent size of the planet's disc or the phase, which are otherwise more or less equal at a given number of days on either side of conjunction. Rather, the answer seems to be simply that observers prefer the convenience of observing Venus in the early evening hours rather than waking up early to spot it before (or after) sunrise. Here we have the second suggestion of an observational bias in historical records of this phenomenon.

More than perhaps any other aspect of the Ashen Light, observers describing it often struggle to find the right words to capture its apparent color as seen through the telescope eyepiece. "Soon after inferior conjunction, when the crescent of Venus is very narrowly illumined, the night side of this planet becomes visible to us in a special shade of color," Schorr wrote. "It is probably impossible to describe exactly this hue, but it almost resembles that of the night side of our Moon when it becomes visible to us." Eyewitnesses through the centuries employed all manner of poetic descriptions in attempting to capture the "impossible" hue. To William Derham, at the dawn of the eighteenth century, it was a "dull and rusty color." In 1866, Friedrich Wilhelm Rudolf Engelmann found it "a lighter, grayish-greenish tone than the background sky." Carl Venceslas Zenger (1830–1908) noted its "peculiar coppery hue" in 1883, while Henry MacEwen described "a beautiful golden light" in 1896. Not to be outdone, the English amateur astronomer Richard Baum wrote first of "a ghastly, gray effulgence," one memorable night in 1953, followed later by "a purplish hue." The disagreement among the reported colors tells us something about either the physical mechanism that produces the light, the human visual perception of color at low light intensities, or both. This is worth a side trip into the strange world of psychophysics, discussed in further detail in Chap. 9.

Finally, there are many instances in which observers report distinct surprise at seeing the Ashen Light at all. In fact, when encountered these accounts tend to jump off the page at the reader. That's particularly true of cases in which the observer already had extensive experience in Venus work, implying many nights on which nothing of the sort was seen. One finds innumerable notes to the effect of "Ashen Light not seen" or "Ashen Light not suspected," clearly indicating that observers were aware of the phenomenon but that they saw on any given night nothing that led them to think they might have detected it.

One may reasonably suspect that some instances of positive detections were caused by observers "primed" to sense something they might believe to be real almost as an article of faith, such that failing to sense the Ashen Light was considered evidence of a defect, real or imagined, in their perceptive abilities. "If one contemplates the few observations of the secondary light of Venus …made in the space of two centuries, it will certainly be admitted that this must be a most rare phenomenon," Schorr wrote in 1875. However:

> if one can deduce the cause of it, however one chooses, then nevertheless several circumstances must co-operate at the same time in order to cause the visibility of the same. However, it can be added that all observers of this phenomenon discovered it only incidentally and were surprised by this perception, initially believing it to be deception. But if they had known beforehand that the night side of Venus must be visible at any time, they would have observed it more often, and we would find a greater number of observations of it recorded.

Primed belief in the existence of the Ashen Light—or, at least, awareness of the sightings claimed by other observers—therefore may not exert an undue influence on the probability that a given Venus observer will see it during his or her career. If the phenomenon is purely psychological, other factors must also be in play.

And what can we make of the very first recorded instance of the Ashen Light, in 1643, by an observer whose perception can't possibly have been influenced by the previous accounts of other people? It's especially noteworthy that the observer carefully noted the presence of an aberration in his instrument that could affect his perception and yet still indicated positively that he had seen something peculiar in regard to many years of Venus observations. Maybe the crescent Venus reminded Riccioli a little too much of the Moon when in its crescent phase, its night side glowing faintly in sunlight reflected from the Earth. Did his brain simply fill in the missing information from an analogous memory?

At the same time, what is the value of *any* particular Ashen Light observation when taken by itself? Roger Cheveau, a French planetary astronomer working in Paris during the interwar period, cautioned about the appraisal of such anecdotes—immediately after reporting his own: "A single observation cannot lead to a serious conclusion on this subject."[18] Yet, still, we have to consider the very real possibility that nothing like a physical "Ashen Light" exists, other than in the minds of those who convinced themselves, one way or another, of its objective reality.

More than 80 years after Cheveau wrote those words, and 400 years after Galileo Galilei became the first human to get a close-up look at Venus, there is still no universally accepted explanation for the Ashen Light. As Julius Benton, Coordinator of the Venus Section of the Association of Lunar and Planetary Observers, put it, "in spite of all the recent arguments against it, the Ashen Light mystery just won't go away," and highly experienced visual observers of Venus insist steadfastly that their impressions of faint light radiating from the night side of Venus are "not illusory."[19] The purpose of the following pages is to examine its history, determine what it is decidedly *not*, pursue some promising leads, and get at the heart of the question of what our eyes really reveal when the brain can very easily tell the heart and mind the wrong story.

[18] 1933, La Lumière Cendrée de Venus, *L'Astronomie*, 47, 69.

[19] 2010, ALPO Observations of Venus During the 2007–2008 Western (Morning) Apparition, *Journal of the Association of Lunar and Planetary Observers*, 53(1), 37.

3

First Light: Early Accounts of the Ashen Light 1643–1800

In the generation after Galileo's pioneering discovery of Venus' phases, early adopters of astronomical telescopes were disappointed when their instruments revealed essentially nothing more about the planet. It remained an impenetrable mystery, a dazzlingly white disc marred by the false rainbow of chromatic aberration. What *was* Venus? Was it just a mirror of the sunlight that illuminated it? What was its nature that made it so reflective? Imperfections in the glass from which telescope optics were then made further distorted planetary images, yielding spurious flares, shadows, and haloes. What was real, and what was not?

In later times, some observers made fantastic claims of dark spots across the Venusian disc, peaks of mountains seen along the boundary between night and day, and even the appearance of a heretofore unknown satellite. As telescopes improved, these reports waxed and waned in frequency not unlike the phases of Venus itself. But a singular account of observations made only 30 years after Galileo pointed his telescope at Venus for the first time sparked a controversy that, unlike other false narratives about the physical nature of Venus, may contain a kernel of truth.

The Observations of Giovanni Riccioli (1643)

Giovanni Battista Riccioli was born in Ferrara, northern Italy, on April 17, 1598. He would later refer to himself in the academic fashion of the times by the Latinized name of his hometown, publishing as "Ricciolus Ferrariensis"

J. C. Barentine, *Mystery of the Ashen Light of Venus*, Astronomers' Universe,
https://doi.org/10.1007/978-3-030-72715-4_3

(Riccioli of Ferrara). While still a teenager, he entered the Society of Jesus for novice training, which he completed 2 years later in 1616. During the same year he began studying humanities at the University of Ferrara, continuing at Piacenza before switching to philosophy and theology at the Jesuit College at Parma in 1620. He ultimately spent 8 years at Parma, during which time he was exposed to contemporary thinking about physics and mathematics.

It was also at Parma that Riccioli first turned his attention skyward. There he learned astronomy under the tutelage of Giuseppe Biancani (1565–1624), one of the most famous Jesuit academics of the day. Biancani is best known for his early telescopic observations of sunspots made in collaboration with Christoph Scheiner (1573/5–1650), a fellow Jesuit astronomer based at the Jesuit College of Ingolstadt in Germany. Riccioli was ordained in 1628 after completing his studies; he then taught first at Parma and later at Bologna. Although Riccioli considered himself primarily a theologian, his avocation remained astronomy until his Jesuit superiors formally assigned him work in astronomical research. After assuming the role of Professor of Theology at the College of St. Lucia in Bologna, he set up an astronomical observatory equipped with state-of-the-art instrumentation, including several Galilean telescopes. His research and publishing in astronomy continued to his death, and he maintained extended correspondence with many famed European astronomers of the day.

Riccioli's magnum opus was *Almagestum novum*, published at Bologna in 1651 (Fig. 3.1). One of the earliest printed books entirely about astronomy, it spanned some 1,500 folio pages in two volumes and quickly became a standard reference work that was still in wide circulation by the middle of the eighteenth century. In it Riccioli related his own telescopic observations of each of the known planets. The description of his Venus work includes a curious account that records details not known to have been witnessed by any telescopic observer before him:

> Although the crescent is a great part of the light of her disc, it is slender, and yields great strength of light when it shines at half-phase. Already, however, we have repeatedly observed Venus in the crescent phase, and more than once in the half-phase, or gibbous, yet it is always blazing: On January 9, 1643, using a telescope that Fontana loaned me,[1] [I observed] the situation and shape [of Venus], which may be seen in the present diagram.[2] It was, compared to the Sun, red to yellowish, and, on the side facing the direction of the Sun, a blue-green: but those colors were most likely caused by the particular variety of glass

[1]Francesco Fontana (*c*. 1580–1656) was an Italian lawyer and an astronomer.

[2]Shown here as Fig. 3.2.

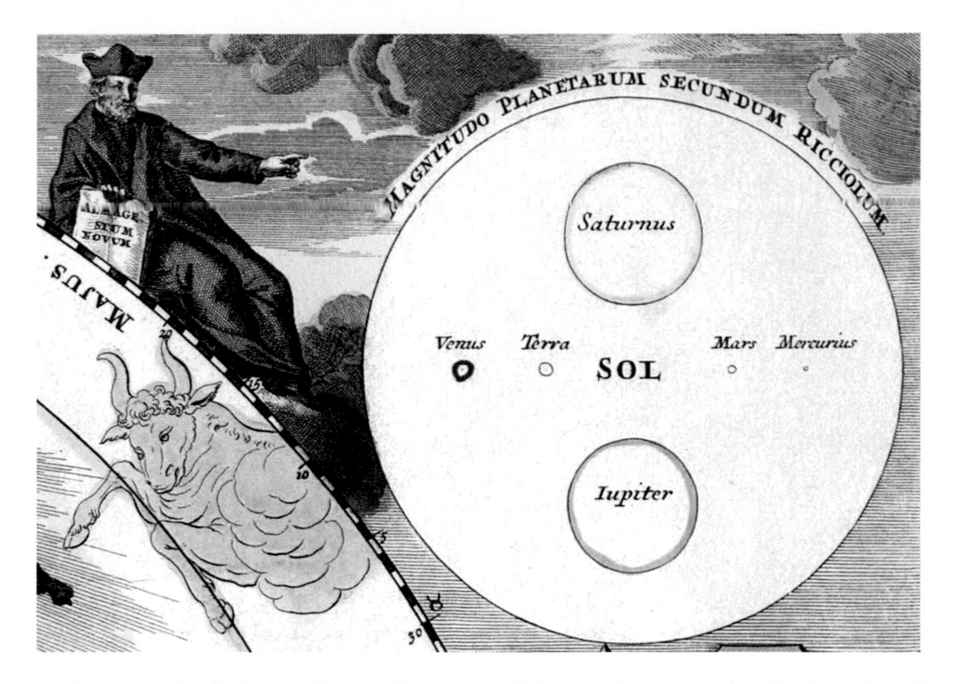

Fig. 3.1 Detail of Plate 3 from Johann Gabriel Doppelmayr's *Atlas Coelestis* (1742) showing Giovanni Battista Riccioli leaning on a copy of his master work, *Almagestum novum* (1651). He points toward a figure labeled Magnitudo Planetarum Secundum Ricciolum ("The Size of the Planets According to Riccioli"), illustrating the supposed diameters of the planets known in his time with respect to each other and to the Sun (outer large circle)

used in the telescope optics.[3] As chance would have it, by Jupiter and Saturn,[4] a partial circle of light could be seen enclosing the otherwise dark side of the disc, brightening it particularly toward the eastern limb.[5]

This wasn't the only instance in which Riccioli thought he saw something anomalous on Venus. Continuing, he relates additional instances:

The same has also been seen at other times, frequently during 1644, when the evening descended, and after rising in the evening, [Venus] became visible near the Sun, and from 7 May until 20 June was seen as gibbous in larger telescopes; [and at] half-phase, around the beginning of July, resembling the Moon at

[3]Riccioli here describes the observed colors of chromatic aberration in an uncorrected refracting telescope.
[4]An oath; compare with 'by Jove' in English.
[5]*Almagestum novum*, Vol. 1, pp. 484–85. Translated by the author from the original Latin.

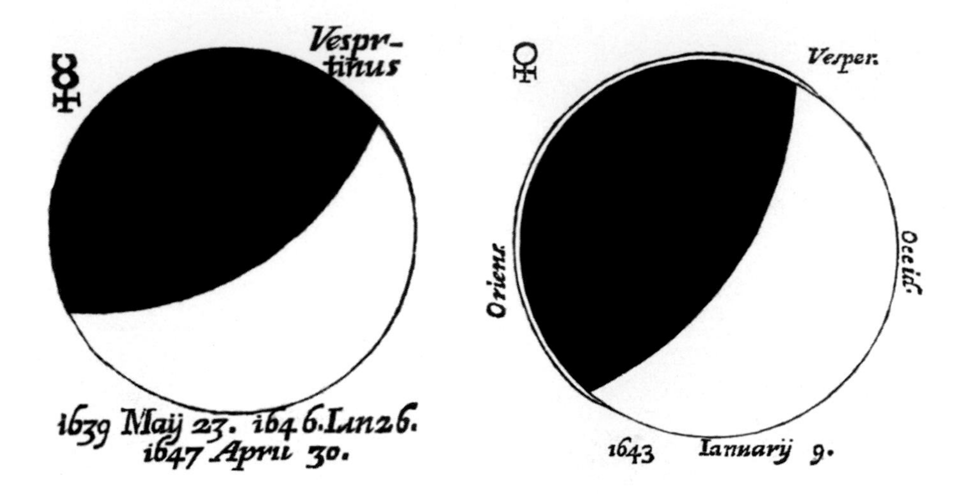

Fig. 3.2 Riccioli's drawings of Venus from *Almagestum novum* (1651). The drawing at right, representing one particular observation and showing an extended "halo" around the night side of Venus, is the earliest known recording of what is presumed to be the Ashen Light in the form of brightening around the limb of the planet's night side (eight o'clock to one o'clock positions)

> quadrature … From that day to 4 August, when we left off observing, it was increasingly sickle-shaped and narrower in width, at length [becoming] more slender, emulating the appearance of the Moon in its first phase: just as it imitates the last phase of the Moon when morning comes with the light of the Sun.[6]

Riccioli indicated that there was something *different* about the appearance of Venus in these cases that he did not observe before 1643 and that it involved the unlit and otherwise invisible side of its disc. His account is generally acknowledged by scholars as the first recorded description of a phenomenon that has since become known as the Ashen Light. Like Galileo's discoveries of the moons of Jupiter, the rings of Saturn, and the phases of Venus itself, this extra light delineating the edge of the Venusian disc against the darkness of the background sky was an unexpected characteristic of what should have been a perfectly bland and uninteresting world in view of the presumed Aristotelian perfection of the heavens.

[6]The "first" and "last" phases of the Moon indicated here are the waxing crescent and waning crescent, respectively.

Early Accounts: 1700–1800

For roughly 70 years after it appeared in print, Riccioli's report was the only one known in surviving records to indicate an unusual appearance to the night side of Venus; we'll find out in the next chapter why that detail is important, because it may tell us something about the nature of what he saw. In fact, his strange description of "a partial circle of light … enclosing the otherwise dark side of the disc" was evidently forgotten for over two centuries.

William Derham (*c.* 1700)

In 1873, Adalbert Safarik (1829–1902), a chemist at the University of Prague, cataloged all of the instances he could find of observers' reports of anomalous light on the night side of Venus up to his time. He presented his findings at a meeting of the British Association for the Advancement of Science, attributing the first observation in history to William Derham (1657–1735; Fig. 3.3), an English theologian and scientist. Derham, he wrote, related an observation he once made:

> in the perigeum of Venus,[7] probably in bright twilight, when he saw the dark side of the planet shining with a dim reddish light. Arago,[8] who mentions this observation, quotes from a French translation published in 1729.

The "French translation" refers to a 1729 edition of Derham's *Astro-Theology: Or a Demonstration of the Being and Attributes of God, from a Survey of the Heavens*, which was first published at London in 1714. This curious book was part 2 of a trilogy whose other titles were *Physico-Theology* (1713) and *Christo-Theology* (1730) and whose intent was to describe a series of teleological arguments for the existence of God and his assumed attributes. Among his musings on the divine, Derham included a number of straightforward scientific observations. *Astro-Theology* announced the discovery of several new "nebulae," a word then used to label any unresolved astronomical object other than a star or planet as seen in the telescope.

[7]An archaic term for inferior conjunction, describing the situation in which Venus appears closest to the Sun and is nearest Earth during its orbit.

[8]Dominique François Jean Arago (1786–1853) was a French polymath and, in later life, President of the Executive Power Commission of the French Second Republic. For 44 days in the summer of 1848, he served as the French head of state.

Fig. 3.3 A portrait of William Derham (1657–1735) from *Crabb's Historical Dictionary* (1825)

In the same year that he published the book, Derham was Rector of the Church of Saint Laurence in Upminster, east London, and had installed a refracting telescope with a focal length of 16 feet in a room at the top of the thirteenth-century church's tower. The intent of the telescope was to assist in the first reasonably accurate determination of the speed of sound in air: Derham used the magnified view to start a timer at the exact instant he saw the flash of a shotgun being fired some known distance away, stopping the timer when he heard the report. While the light travel time was essentially instantaneous, a delay of several seconds in the arrival of the sound allowed its velocity to be computed in a straightforward way.

By night Derham used the telescope to observe the heavens. In *Astro-Theology*, he reported an observation of Venus that by turns seems both incidental and extraordinary:

> At such times as these Planets shew their full Phases, they are found to be spherical, and only lose this Figure by virtue of their Position to the Sun, to whom they owe their Light. And this Sphericity, or Rotundity is manifest in our Moon, yea and in Venus too; in whose greatest Falcations[9] the dark Part of their

[9]An obsolete term for the crescent phases of the Moon and planets, derived from the Latin word *falx*, referring to the shape of a sickle or scythe.

> Globes may be perceived, exhibiting themselves under the Appearance of a dull, and rusty colour.

Derham indicates here a proper understanding of the telescopic appearance of the "inferior" planets Mercury and Venus, whose orbits inside that of Earth cause them to undergo a cycle of phases much like those of the Moon. This is unlike the other, "superior" planets orbiting beyond Earth whose discs only ever show the slightest degree of phase and are otherwise fully round in appearance. In referencing the thin crescents of Mercury and Venus near inferior conjunction—in their "Falcations"—Derham at first seems to mean that what he saw on Venus was like the "earthshine" seen on the otherwise unlit portion of the Earth's Moon in its own thinner crescent phases.[10] As a straight piece of reporting unaffected by interpretation, it's not unreasonable to use the description Derham did, even if the underlying phenomenon isn't the same.

However, in a footnote Derham specifically qualifies what he saw as the "secondary light" of Venus and gives a more detailed description of his telescopic observation:

> What I have here affirmed of the Secondary Light of Venus, I have been called to an account for, by an ingenious Astronomer of my Acquaintance. But I particularly remember, that as I was viewing Venus some Years ago, with a good 34 foot Glass, when she was in her Perigee, and much horned, that I could see the darkened Part of her Globe, as we do that of the Moon soon after her Change. And imagining that in the last total Eclipse of the Sun,[11] that the same might be discerned, I desired a very curious Observer that was with me, and look'd through an excellent Glass, to take Notice of it, who affirmed that he saw it very plainly.

Derham's account has since been interpreted as an unusual and early report of the Ashen Light under circumstances that rendered the disc of Venus, which was then very near the Sun, unusually large, and yet with a comparatively dark night sky behind it. As we'll see later, there have been many other accounts of observers who said they saw something unexpected about the unlit side of Venus even during observations made in full daylight, but Derham's report is

[10] See Chap. 5 for a full exploration of this phenomenon.

[11] The German astronomer Albert Marth (1828–1897) determined the date of the eclipse to which Derham refers here as May 3, 1715, N.S. (footnote on page 111 in *The Astronomical Register* Vol. 14, No. 161, May 1876). The date of this eclipse helps constrain the time period in which Derham made his first Ashen Light observation "some years ago."

unique as probably the earliest instance of an Ashen Light observation made during the circumstances of a total solar eclipse. We don't know exactly when Derham first saw the effect that he likened to earthshine, but "some years" prior to 1715 puts the observation somewhere around the year 1700. Derham's account is also the first of which we know where more than one individual (in this case, his "very curious observer" friend) saw the effect under the same circumstances on a given night.

His "34 foot glass" was a so-called aerial telescope, and its description gives us some useful information for comparing Derham's observation to Riccioli's view. In the mid-seventeenth century, astronomers found that if they ground the glass of their telescope lenses to have very long focal lengths, the effects of chromatic aberration could be diminished or even eliminated. In contrast to the telescope Riccioli used, Derham's telescope was probably sufficiently free of chromatic aberration that his description of a "rusty color" on the night side of Venus becomes more believable.

As of about 1720, the surviving historical record appears to contain only these two observations, one of which was singular in the sense that Riccioli noticed something different or unusual about Venus on certain nights that he didn't see on others and for which he had no point of reference or comparison. Derham, who makes no mention of Riccioli and hence likely had no prior knowledge of the earlier observation, ventured a more specific explanation for what he saw, like earthshine giving the night side of Venus a faint cast not unlike the otherwise dark part of a crescent Moon.

These accounts were largely unknown in Europe during the mid-eighteenth century, and it is likely that essentially no one was routinely observing Venus for the purpose of trying to detect any faint, night-side light. For the better part of 100 years, their observations represented the entirety of what anyone knew about the Ashen Light. Still, these two eyewitnesses are individually short of what we might want in terms of reliability: Riccioli was experienced as an observer but lacked a good instrument, while Derham had an arguably better telescope but spent less time routinely at the eyepiece. History's next recorded observation was made by a figure who had both qualities.

Christfried Kirch (1721)

Christfried Kirch (1694–1740) was a German astronomer literally raised in the craft, the son of Gottfried Kirch (1639–1710), the first German "astronomer royal" and first Director of the Berlin Observatory, and Maria Winckelmann Kirch (1670–1720), an astronomer in her own right and one of the first women in history to achieve wide recognition and fame for her work in the field. From his childhood, Christfried assisted his father and mother in their observations, and he eventually succeeded his father as Director of the Berlin Observatory from 1716 until his death.

He made many planetary observations during his life and twice reported the Ashen Light,[12] according to Safarik "with moderate optical power and in bright twilight." Kirch wrote in his observing notebook:

> 1721 June 7. Saturday evening in the Observatory, I found Venus in a part of the sky that was not particularly clear...She was very narrow (according to a measure of the diameter = 1′5″). It seemed to me as if I saw the dark part of Venus, which strikes me as rather incredible. The bright part [of the disc] was roiling because of the vapors.[13]

Kirch's expression of surprise at apparently seeing the unlit part of the disc of Venus appears to be history's first instance of scientific skepticism of the Ashen Light, which continues unabated into our own time.

To his evident surprise, Kirch again saw faint light from the night side of Venus a few years later:

> 1726 March 8. Friday...Then we went upstairs and observed the crescent Venus with the 26-foot telescope. I guessed the illuminated part of Venus was some arcseconds wide... I could see the dark part of Venus, and that dark outer circumference appeared to me to be of a somewhat smaller circle than the inside, but as usually happens with the Moon when the unlit part can be seen, the cause is that the bright light [from the lit side] is propagated in our eyes, and seems bigger than it is in fact.[14] Mr. Harper and Mr. Möller assured [me] also that they recognized the dark part. It seemed to me in this way before (followed by a slight sketch). Note that I used the usual aperture of the telescope, and not with an aperture stop, as one is wont to use otherwise when observing Venus.

Kirch's last comment here is of particular interest, as it gives us some idea about *how* planetary observations were typically made in an era when such

[12]Recounted by Eduard Schönfeld, 1866, *Astronomische Nachrichten*, 1586, 27.

[13]Here Kirch refers to the distorting effect of turbulence in the Earth's atmosphere, through which light from astronomical objects passes before reaching the telescope.

[14]Kirch correctly understood what he saw as an effect of the eye-brain combination. Just as bright stars are apparently slightly larger to the naked eye than faint stars, the brightness of the lit part of the crescent Moon in relation to the strong contrast with the much lower-intensity night side, including earthshine, leads the brain to believe that the day side is disproportionately large compared to the size of the rest of the disc.

work was dominated by the use of progressively larger, long-focus refracting telescopes. These instruments were typically fitted with adjustable irises that allowed the aperture to be varied from the full diameter of the objective lens down to any arbitrary size chosen by the user. While large telescope apertures collect a lot of light, helping keep planetary images bright even at high magnification, they suffer disproportionately from the blurring effects of the Earth's atmosphere. Trial and error demonstrated the existence of a "sweet spot" of apertures of about 6 to 10 inches that offered the best compromise between sufficient light-gathering power and beating the blurring effects of the atmosphere, defying the naïve notion that bigger is always better when it comes to astronomical telescopes. Visual observers of the planets to this day play a game in which they try to balance the competing needs of light-gathering capacity, angular resolution, image stability, and visual contrast in attempting to sense fine details on planetary discs.

Kirch, like many modern Venus observers, was otherwise inclined to employ an aperture stop to both improve seeing conditions and reduce the intensity of the dazzling Venusian disc, helping to draw out faint details seen by observers as early as the mid-seventeenth century. That Kirch states he did *not* make use of an aperture stop in the 1726 observations works for and against the verisimilitude of an Ashen Light phenomenon on the Venus night side: the extra aperture may have increased Kirch's ability to sense faint light on the night side of the planet, but it also admits more of the day-side light whose scattering in the optical system of the telescope could be mistaken for something real on Venus. Unfortunately, Kirch left no comment in his observing notebook as to whether he tried stopping down the aperture at any point during the episode or if he still saw light from the night side of the planet after doing so.

Andreas Mayer (1759)

Thirty years later, another German astronomer, Andreas Mayer (1716–1782), saw something that reminded him of earthshine on the Moon, but rather on the night side of the crescent Venus. Mayer was initially trained in land surveying and architecture by his father of the same name, later studying astronomy with Christfried Kirch at Berlin before receiving a doctorate in mathematics from the University of Wittenberg in 1736. Five years later he was appointed Professor of Mathematics and Astronomy at the University of Greifswald in what was then Swedish Pomerania, a Swedish dominion on the Baltic Sea coast of Germany and Poland, where he remained until his death.

Mayer set up a small observatory at the university from which he recorded a curious account of Venus in full daylight at 44 minutes past noon local time on Saturday, October 20, 1759. Adalbert Safarik later wrote that Mayer spotted Venus "only 10° from the sun with an unachromatic transit-instrument of only 1 1/2-inch aperture."[15] Mayer himself wrote that "although the illuminated portion of Venus was quite small, still the entire disc appeared, in the likeness of the crescent moon, which reflects the light received from the Earth," indicating the perception of an effect similar to earthshine.[16] The unilluminated portion of the disc seemed to Mayer distinctly *brighter* than the daytime sky around it.

It was believed for many years that Mayer's daytime observation was singular in history, and decades again passed before anyone who saw the Ashen Light seems to have commented about it in print. Meanwhile, both telescope technology and the quality of the glass from which refractor objectives were made improved steadily during the eighteenth century and removed the plague of chromatic aberration from which early refracting telescopes suffered. Some astronomers simply bought or built reflecting telescopes, whose designs rendered them intrinsically free of spurious color defects but required regular maintenance. At the turn of the century, telescope mirrors were cast from "speculum metal," an alloy of roughly two-thirds copper and one-third tin that resulted in a brittle, white metal that could be polished to high reflectivity. Unfortunately, these mirrors tarnished quickly once exposed to typically humid European air, and they required frequent polishing to keep their shine. Diligence was, however, in many cases, rewarded.

William Herschel (1793)

William Herschel (1738–1822) apparently made the first recorded observations of the Ashen Light with a reflecting telescope. Born into a family of musicians in the Electorate of Hanover in what is now Germany, Herschel came to England at the age of 19 while in the military service of Hanover, which at the time was ruled in personal union with the Kingdom of Great Britain. Herschel went on to become one of the most prolific observational astronomers of his time, cataloguing with his sister, Caroline (1750–1848), the positions of thousands of double stars, nebulae, and other night sky objects. He was also

[15] Here Safarik means that the telescope Mayer used was not corrected for chromatic aberration, typical of certain small telescopes used to make astronomical timing measurements.

[16] 1762, *Observationes Veneris Gryphiswaldenses*, Greifswald: A. F. Röse, 19. The observation is recounted in Johann Hieronymus Schröter, 1811, *Beobachtungen des grossen Kometen von 1807*, Göttingen, Appendix, 74.

a lifelong observer of the planets and achieved infamy when he became the first person since antiquity to discover a planet, Uranus, in 1781. With his telescopes, some of which were the largest on Earth at the time, Herschel was the first to spot several moons of Saturn and Uranus and the first to note the seasonal changes in appearance of the polar ice caps on Mars. While remembered principally for his observations and discoveries in the "deep sky," Herschel is rightly considered an important contributor to early planetary astronomy.

The Venus program Herschel undertook over several decades involved a search for the appearance of clouds or mountains, both reported previously by other observers, but he ultimately found no convincing evidence of either. However, he noted the existence of an effect on Venus analogous to twilight on Earth, in which the atmosphere scatters sunlight from the day side around to the night side of the planet. As a result, the line of the "terminator," separating light from dark, is not as sharply defined as it is on the effectively airless Moon. Rather, on a body with a significant atmosphere, the terminator is fuzzy and indistinct; twilight lengthens the transition from day to night, softening and filling in shadows cast by objects on the surface.

In May 1793, Herschel recorded an apparent extension of the "horns" of the crescent Venus beyond expectations if no atmospheric scattering were involved, adding evidence for a Cytherean twilight. "The atmosphere of Venus is probably very considerable," he wrote, "which appears not only from the changes that have been observed in the faint spots on its surface but may also be inferred from the illumination of the cusps, when this planet is near its inferior conjunction, where the enlightened ends of the horns reach far beyond a semicircle."

Unlike the Ashen Light, the mechanism by which the horns of the Venus crescent appear elongated is well-understood, and Herschel knew the right answer. But the light scattering effect is so strongly biased in the forward direction that the horn extensions only appear within a few days on either side of inferior conjunction (Fig. 3.4).[17] Herschel was therefore perplexed when he saw something reminiscent of the effect a full *7 weeks* before inferior conjunction on May 27.

[17] "As Venus approaches within 15° of the Sun, faint luminous arcs creep out from the needlelike cusps to extend the slender crescent beyond its geometric limit. … These arcs continue to advance around the dark rim of the planet until finally at inferior conjunction they encircle Venus with a ring of faint bluish light. The effect is at once dramatic and spectacular." (R. Baum, *Journal of the British Astronomical Association*, Vol. 105, No. 5, p. 216, October 1995). Baum warned that the observation of this phenomenon through a telescope is literally so dangerous, due to the inevitable proximity of the Sun, that it "*is not for the inexperienced.*" (emphasis in the original).

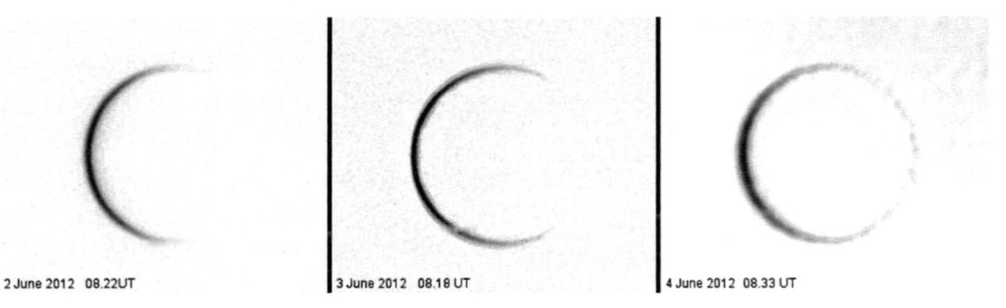

Fig. 3.4 An example of "cusp extensions" of the Venus crescent, imaged by John Sussenbach in the Netherlands in the days leading up to the inferior conjunction (and Venus transit) of June 5, 2012. Strong forward scattering of light in the Venus atmosphere is evident; on the day before the 2012 transit (right), the entire circle of the planet's disc is outlined in light. The grayscale images have been inverted for better clarity (Used with kind permission of the photographer)

In a report to the Royal Society later that year, Herschel stated that he observed Venus on the night of April 9 "with the 10 feet reflector, power 300," finding that "the light of Venus is brighter all around the limb, than on that part which divided the enlightened from the unenlightened part of the disk." But on the same night he mentions having seen something remarkably like that first described by Riccioli in 1643: "The bright part, on the limb of Venus, is like a bright bead, of nearly an equal breadth all around." It appeared again on April 16 as "a luminous margin, as usual, all around the limb," suggesting that the glow at the edge of the night side limb was a recurrent phenomenon over multiple nights. That he mentioned it at all is noteworthy: it was clearly unlike anything he had seen among his observations of Venus, which at that point reached back two decades.

Four nights later Herschel saw the same apparent effect, which he described in more detail. Again, though the 10 foot reflector:

> with 287 [magnifying power], there is a narrow luminous border all around the limb, and the light afterwards diminishes pretty suddenly, and suffers no considerable diminution as we go towards the line which terminates the enlightened part of the disk. It is however less bright near the terminating line than farther from it. With powers lower than 187, the narrow luminous border cannot be so well distinguished.

And again on April 22, "with 430 [magnifying power], the luminous margin, compared to the light adjoining to it, may be expressed by, *suddenly much brighter all around the limb*." (emphasis in the original)

In his Royal Society paper, Herschel differentiated between the twilight phenomenon and the glow he described as "like a bright bead, of nearly an equal breadth all around" the limb of the night side of Venus:

> The remarkable phænomenon of the bright margin of Venus, I find, has not been noticed by the author we have referred to:[18] on the contrary, it is said, page 310, "this light appears strongest at the outward limb…, from whence it decreases gradually, and in a regular progression, toward the interior edge, or terminator." But the luminous border, as I have described it, in the observations of the 9th, 16th, 20th and 22nd of April, does not in the least agree with the above representation.

Based on what he saw, Herschel put forth an explanation for the Ashen Light in the context of limb brightening that brought a modicum of scientific reasoning to the discussion:

> With regard to the cause of this appearance, I believe that I may venture to ascribe it to the atmosphere of Venus, which, like our own, is probably replete with matter that reflects and refracts light copiously in all directions. Therefore on the border, where we have an oblique view of it, there will of consequence be an increase of this luminous appearance.

It was a plausible idea for why the ghostly glow suddenly became much more apparent at the limb of the planet, but unlike what is observed on Earth it required that sunlight was conducted far beyond local midnight on Venus. If true, it implied that the atmosphere of Venus was not at all like our own.

Friedrich von Hahn (1793)

Adalbert Safarik summarized the extensive observations of the Ashen Light made around the same time by Count Friedrich von Hahn (1742–1805), a nobleman and astronomer working in Mecklenburg, a historical region in northern Germany. From his private observatory in the village of Remplin, von Hahn:

---

[18]This appears to be Johann Hieronymus Schröter, based on Herschel's reference in another footnote to *Beiträge Zu Den Neuesten Astronomischen Entdeckungen* (1788) by Schröter and Johann Elert Bode, and Schröter's *Selenotopographische Fragmente* (1791). Schröter himself first reported seeing the Ashen Light during observations of Venus on May 21, 1793, according to Schorr [64]. Herschel here makes a point that his observations did *not* qualitatively agree with Schröter's.

saw the phenomenon unusually well and often during the spring and summer of 1793, in twilight as well as in daylight. He employed excellent instruments, and gives a very detailed description of what he saw; also two sketches. No other observer seems to have seen the phenomenon so often and so well.[19]

von Hahn's observations were noted in a section of the *Berliner astronomisches Jahrbuch für 1796* entitled "Bemerkungen an der Venus, Beschreibung einiger merkwürdigen Sonnenflecke und astronomische Nachrichten" ("Comments on Venus, description of some strange sunspots and astronomical news".) He began by noting his frequent observations, including those made in the daytime:

> For some time I have frequently observed Venus, even specifically in daylight. … Often I saw with great clarity the dark part of the disc of the planet, made remarkable by a gray or faded brownish color. I would like to know if any spurious light or similar cause could explain this phenomenon, which is perceived by [observers using] several telescopes if the air is clear. … Sometimes I believed the [disc shape] completely rounded, surrounded by a fine circle of light at the edge of the disc, seen before me.

von Hahn didn't know what to make of what he saw, nor did he evidently care to make a guess. "I sometimes saw very clearly the unenlightened part of the disk, but I myself abstain at present from all hypotheses," he wrote. "This perceived appearance by several excellent telescopes is often regarded with astonishment. It may be that still an illusion prevails." He implied that the "illusion" is one of scattered light from the "the circle of confusion of the vibrant rays of Venus." von Hahn seems to have been the first writer to suggest that the Ashen Light might simply be an illusion of the telescope, eye and brain, indicating nothing of a physically luminescent Venusian atmosphere. Indeed, his illustrations of what he saw through his telescope (Fig. 3.5) suggest that the extent to which the night side seemed apparent depended on the phase of the planet, decreasing in proportion to the changing illumination of the day side. Such behavior is broadly consistent with light from the bright crescent scattered into the darkness beyond.

By about 1800, the Ashen Light remained something of a fashionable mystery, treated by most astronomers as real but rare. Published accounts

[19] Safarik cites his source as "*Berliner astronomisches Jahrbuch fur 1793*, p. 188," but there is no mention of Venus on that page. Rather, it appears that Safarik quoted the wrong *Jahrbuch*. A publication for a given year documented events of the 3 years previous; the correct reference is the same page number in the *Jahrbuch* for 1796.

Fig. 3.5 Figures 4 and 5 from Friedrich von Hahn's report in the *Berliner astronomisches Jahrbuch für 1796*. von Hahn described observations in Venus during 1793 in which the night side was visible and hence "replaced by an oval shape" in Fig. 4. However, the appearance of the night side was diminished as Venus approached inferior conjunction on May 27: "Now, during the planet's crescent phase, this is no longer quite as conspicuous. By contrast it is is now bounded with a black stripe, as Fig. 5 shows"

in the latter half of the eighteenth century increasingly referenced earlier reports, supporting the conclusion that it was at least generally known among observers, even those who had never seen it. It was also the close of a period of history in which astronomers treated observational work as a form of stamp collecting and the beginning of a time when they began to attempt to make physical sense of what they were seeing.

4

Going Mainstream: A Scientific Approach c. 1800–1900

Nothing quite like the modern science of astrophysics existed at the time Giovanni Riccioli made his observations of Venus in the first half of the seventeenth century. Even after the introduction of the telescope in the same era, the study of astronomy was largely one of phenomena only loosely attached to concepts of physics yet still firmly associated with the pseudoscience of astrology. Astronomers of the late pre-telescopic era such as Tycho Brahe (1546–1601) made precise observations of the positions of the planets that were used as inputs to mathematical models of their positions, but the models largely existed only to validate the preexisting belief in an Earth-centered Solar System. As it turns out, the seeds of geocentrism's undoing were already sown in Brahe's measurements. It just took an old idea to make them grow and bloom.

In the third century BCE, Archimedes of Syracuse attempted to constrain the size of the universe by trying to figure out how many sand grains could fit inside of it. But in the Hellenistic era in which Archimedes lived, there were no words in the Greek language to properly express numbers of the size required to describe something as large as all that ever existed. Greek words then could name numbers up to a "myriad," representing 10,000 or ten to the fourth power (10^4). The myriad itself could in principle be used as a unit, and counting could then proceed up to a myriad myriad (10^8). Archimedes carried this process out recursively until he arrived at the largest number the system imagined: a one followed by some eighty quadrillion (8×10^{16}) zeroes.

Archimedes needed access to notation that could handle computations dealing in such large numbers because he had reason to believe that the

J. C. Barentine, *Mystery of the Ashen Light of Venus*, Astronomers' Universe,
https://doi.org/10.1007/978-3-030-72715-4_4

universe was a *very* big place. In *The Sand Reckoner*, he made reference to a now-lost work of his contemporary, Aristarchus of Samos, who:

> brought out a book consisting of some hypotheses, in which the premisses lead to the result that the universe is many times greater than that now so called. His hypotheses are that the fixed stars and the sun remain unmoved, that the earth revolves about the sun in the circumference of a circle, the sun lying in the middle of the orbit, and that the sphere of the fixed stars, situated about the same centre as the sun, is so great that the circle in which he supposes the earth to revolve bears such a proportion to the distance of the fixed stars as the centre of the sphere bears to its surface.[1]

Although Archimedes grossly underestimated the size of the universe at 10^{14} Greek *stadia*, or about 1.85×10^{13} km, the thinking that led him to the result was revolutionary. In fact, given the exposition and analysis of its topic, *The Sand Reckoner* is sometimes regarded as the first scientific paper in history. This is made all the more dramatic by the fact that in it Archimedes accepts the Sun-centered Solar System of Aristarchus as one of the core assumptions in his chain of logical steps. In other words, it was treated in its time as a noncontroversial conjecture rather than itself a result.

It was a development almost as big as the fantastically large numbers he calculated, but the decision to adopt the new model of Aristarchus was more a matter of pragmatism than conviction on Archimedes's part. The Greeks tried and failed to observe the "parallax" of the stars, which created a problem for the geocentric theory. In the traditional construction of the universe with the Earth at the center, the stars were immovably fixed to the surface of a sphere just beyond the orbit of Saturn, then known as the outermost of the planets. When observed from widely separated locations on the Earth, even Saturn showed a slight parallax or angular shift with respect to the stars; this is the vertex angle of a long and very skinny isosceles triangle whose base is the distance between the observation points on Earth and whose height is the distance from the Earth to Saturn. The Saturn parallax was a tiny angle but one measurable using contemporary techniques. Being only a little further away, the stars should also demonstrate a parallax. Yet none was ever seen, implying that they were instead much further away than Saturn.

Archimedes needed a large universe for his calculation, evidently because he felt innately that it must be much larger than was believed up to his time. The Aristarchan universe fit the bill, for a heliocentric model didn't assume that the

[1] Translation of Thomas L. Heath, Cambridge University Press, 1897, p. 520.

stars were at any particular distance from the Earth. Almost as an aside, it also solved the apparent problem of the missing stellar parallaxes. We don't know how this approach was received in Archimedes's time, but we do know that, like much of the scientific literature of the Hellenistic era, it disappeared from view for nearly 2,000 years.

Whether or not he was aware of Archimedes or Aristarchus, the Polish mathematician and astronomer Mikołaj Kopernik (1473–1543)—better known by the Latinized form of his name, Nicolaus Copernicus—proposed the same heliocentric theory around 1510 on the basis that it was fundamentally simpler and more elegant than geocentrism, requiring fewer corrections in order to fit the observed motions of the planets. Famously, legend has it that Copernicus finally saw his idea in print on his literal deathbed, receiving the printer's proofs of his major treatise, *De revolutionibus orbium coelestium*, on the very day he died.

Johannes Kepler, working almost a century after Copernicus, made a careful study of Brahe's many years' worth of naked-eye observations of planetary positions. Using the data, he further refined the Sun-centered model and from it extracted a series of laws of planetary motion that yielded the first hints of the underlying mechanics. But it was the English polymath Isaac Newton (1643–1727) who first applied physical reasoning to arrive at the correct answer: not only do the planets revolve around the Sun and not the Earth, but also the force of gravity explains their motion and can be used to accurately predict their future positions in the night sky.

Prior to the emergence of the scientific method and the physics needed to explain what humans saw when they looked out into the cosmos, astronomy was a largely phenomenological discipline: its practitioners could describe what they saw, but they really couldn't say *why* any of it was as it was seen. The transition to the predominance of astrophysics we know today began in the Enlightenment and was more or less complete as the public investment in science ramped up quickly in the West during the economic expansion following the end of World War II. The process that took two centuries to complete involved quantitative measurements displacing qualitative observations as the physics required to understand the numbers matured. The last major missing piece of the puzzle, quantum theory, was elucidated in the early twentieth century, providing the conceptual framework to finally understand fundamental astrophysics such as how stars make light. While there will always be frontiers to explore, humans now know a great deal about how the universe works.

By the turn of the nineteenth century, astronomers in Europe and North America increasingly sought to fit their observations into the context of physical theories that offered genuine insight into the nature of the universe. But in Riccioli's time, an observation of Venus like the one he recorded in *Almagestum novum* was cause for curiosity and sometimes wild speculation; while the telescope brought phenomena apparently *closer* to the human eye and brain, it didn't yet add any specific information to our understanding of the cosmos. That required time in the laboratory in order to understand physics, as well as the development of technology that could both sense light more efficiently than the human eye and extract information from that light beyond characteristics such as intensity and color.

Meanwhile, the Industrial Revolution was making goods of all kinds easier and cheaper to produce. Telescopes could now be built largely from machine-made parts, but the production of fine optics required the skilled hand of the master to perfect. Although still only accessible to the wealthy, more people got their hands on instruments suitable for making meaningful observational contributions to astronomy, installing them on rooftops and in back gardens. Nearly any university with access to modest purchasing resources could acquire a small refracting telescope and outfit an observatory for the benefit of its astronomers, and astronomers in turn relied less on the largesse of individual patrons to enable their observational programs. Astronomy slowly—*very* slowly—grew slightly more democratized and inclusive as it became vastly more scientific.

The Moon and planets were objects of intense scrutiny, given that they were relatively nearby and therefore bright, making them readily accessible to smaller apertures; by comparison, other than meticulously measuring the delicate dance of orbiting binaries, the stars gave up no secrets through the eyepiece. The rest of the universe was a complete mystery. But the average observer with a small telescope could map lunar features, make timings of the bright moons of Jupiter, and sketch shadowy markings on the globe of Mars as it rotated with a remarkably earthlike 24-hour period. Speculation continued to dominate astronomical thinking, but that speculation began to collide with skepticism brought about by an increasingly empirical attitude that demanded evidence as the best way to figure out how things actually worked.

The proliferation of telescopes in the nineteenth century, and the tendency of their users to focus them on the most conveniently obtainable objects, meant that the number of people routinely observing Venus grew at a healthy rate. More eyeballs on Venus more of the time implies a higher probability that any given observation would yield a report of the Ashen Light; indeed, reports saw a bump in the latter half of the century (see Fig. 2.7). But it is also clear from observers' accounts that, in many cases, they questioned even their own senses in deciding whether what they perceived was real in the first place.

Johann Hieronymus Schröter and Karl Ludwig Harding (1806)

Johann Hieronymus Schröter (1745–1816; Fig. 4.1) was a German lawyer trained at Göttingen who in 1777 found himself appointed Secretary of the Royal Chamber in Hanover to King George III of the United Kingdom, who concurrently ruled Hanover as its Duke and Prince-Elector. While at court, Schröter made the acquaintance of two of William Herschel's brothers and ultimately that of the astronomer himself. Herschel's discovery of the planet Uranus in 1781 inspired Schröter to spend more of his time in the study of astronomy, and shortly thereafter he resigned his post at Hanover to take a new assignment as the chief magistrate and district governor of Lilienthal in Lower Saxony, where he remained to the end of his life.

Schröter's second career as an astronomer bloomed in Lilienthal. He acquired a series of successively larger reflecting telescopes in the 1780s (Fig. 4.2), which he used in a campaign of systematic observations of Mars, Venus, Jupiter, and Saturn. Schröter was, however, best known for his lunar work, which yielded one of the first comprehensive lunar studies ever published, *Selenotopographische Fragmente zur genauern Kenntniss der Mondfläche* ("Topographical scraps for a more precise knowledge of the lunar

Fig. 4.1 Contemporary portrait engravings of Johann Hieronymus Schröter (left) and Karl Ludwig Harding (right)

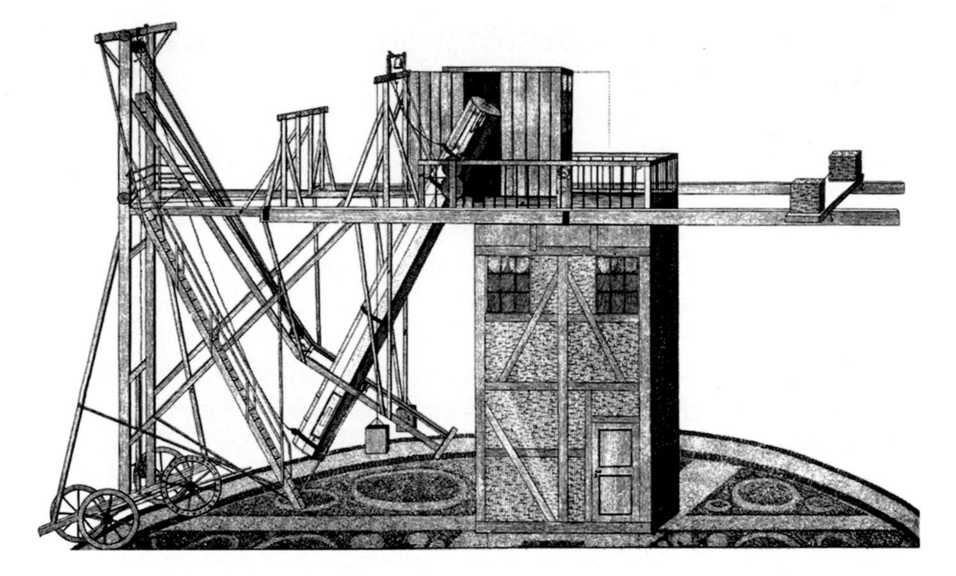

Fig. 4.2 An engraving after Schröter's own 1793 drawing of the 27-foot telescope at his "Second Observatory" in Lilienthal, Germany, with which he made observations of the Ashen Light in February 1806. Adapted from Plate 1 of the 1796 edition of Schröter's *Aphroditographische Fragmente*

surface," 1791). Later in life he fell victim to collateral damage from the Napoleonic Wars, his observatory, its equipment, and many of his journals destroyed in 1813 by French troops. He died a broken man, his enterprise in ruins.

In 1796 Schröter hired Karl Ludwig Harding (1765–1834) to tutor his son. Harding was also educated at Göttingen, in theology and natural philosophy, and harbored his own interest in astronomy. After a few years, Schröter brought Harding on as an assistant at his observatory; while there, Harding discovered (3) Juno, then only the third-known member of the asteroids, a new class of objects whose very existence flummoxed the best minds of Europe. Later he returned to Göttingen as the assistant of Johann Carl Friedrich Gauss (1777–1855), one of the most prolific and influential mathematicians of all time. Harding became a professor of astronomy there and published *Atlas novus coelestis* (1823), a catalog of some 120,000 stars.

In early 1806, shortly after Harding departed the observatory for his new post in Göttingen, Schröter was in the midst of routine observations of Venus when he noticed something odd about the planet. "At 7pm on February 14, 1806, the air was in turmoil," he wrote:

> and Venus stood only about 10 degrees high. But this notwithstanding, she appeared, with 150 times magnification and with the full aperture of the exquisite 15-foot reflector, delimited extremely sharply and beautifully in her crescent shape, and both horns were equally regular and pointed. But without thinking of anything, … and with the 27-foot reflector stopped to an aperture of 20 inches, observations of the crescent-shaped figure [showed], for the first time, the whole of the planet in its atmospheric twilight, not illuminated by the sun, in its nightly form, [but rather] in an extremely dull, dark light.

Even though this "dull, dark light" took Schröter by surprise and caused in him some skepticism, he somehow convinced himself of its reality. "It was not, however, illusion," he asserted, "because I saw it against the strong illumination of the day side just as noticeably strong as around its circumference, as in the case of the unilluminated portion of the crescent moon to the unaided eye, and because I could not easily explain it away as an optical illusion, I would like to imagine it as I wanted."

The following night was "too hazy" to observe Venus, Schröter noted in his account, and the subsequent 6 days were "completely covered with clouds." His next view of the planet came on February 21 and suffered some interference from the nearby waxing crescent Moon, but on that night he detected nothing out of the ordinary: "I could not see the same phenomenon again with the same telescope, even with the sharpest seeing." The same null result was obtained during the next series of clear nights (February 23, 24, and 28). He continued, with emphasis in the original:

> This proves that the only possible observations [of this phenomenon] are limited to relatively few days around the crescent phase of Venus[2] *under otherwise favorable conditions*, and that even for the most diligent observer this is therefore a coincidence, because twilight and the very dim light of the night side of Venus do not permit such an appearance. It is therefore limited only to certain elongation angles of the planet; and it is also possible that it may be even rarer if, perhaps, during an apparition of Venus, one hemisphere of the Earth is somehow capable of reflecting more light than the other.

Schröter believed his observation was the first in which diffuse light was seen across the entirety of the night side of the Venusian disc: "Admittedly Dr. Herschel, Marshal [Friedrich] von Hahn and myself saw the effect first somewhat more on the tips of the horns, and sometimes less on the night side

[2] Schröter was observing some 2 to 3 weeks before inferior conjunction, which occurred on March 12.

of Venus, but, as far as I know, the appearance of the whole dark sphere has not yet occurred to any single observer." This suggested to Schröter something more than just a simple "extended twilight" from a dense atmosphere was involved, implying that the source of the faint glow he saw was not scattered light from the Sun. Furthermore, he reasoned, if the explanation for what he saw involved typical conditions on Venus, then it should be visible all the time. Yet the glow that Schröter saw was clearly not typical to him.

Two hundred kilometers away in Göttingen, Harding was also observing Venus during its evening apparition in early 1806, and he recorded what he saw on nights adjacent to those mentioned in Schröter's report. Schröter wrote to Harding about his observations after the fact, noting later that "this excellent observer responded to my delight that he perceived exactly the same phenomenon already on 24 January." In claiming his own independent detection of the dull light of the Venus night side, Schröter concluded, Harding "confirmed the truth" of the earlier observations of Herschel and von Hahn.

Concerning what he saw, Harding's pronouncement was even stronger than his former mentor's: it was, he reported in the same edition of the *Berliner Astronomisches Jahrbuch*, "the entire disc of Venus before me, just like the Moon." Again with emphasis in the original, Harding related that:

> when I saw Venus with the 10-foot Herschel Reflector on January 24, in misty but very pure air, I noticed immediately at 84 times magnification and with the full opening of this beautiful instrument, the entire planet's face, *not illuminated by the Sun*, which was clearly marked by a dull, ashy light against the dark sky. At first I was inclined to regard this phenomenon as a deception, since it would have been possible for it to be the work of the imagination, which, Venus being just like the Moon, just before and after new, and where, for known reasons, it is also seen on the night side, it is easy to comprehend the phenomenon always accompanying it in this form.

The emphatic nature of Harding's words is probably what moved Schröter to the belief that this was indeed a powerful affirmation of his own impressions. As much as Harding claimed to not be predisposed to believe what he saw, "the sight was so clear that no conception of the contrary could obscure it, and as it remained the same in all points of the visual field, I could judge nothing other than that the appearance must be real."

Venus remained the object of Harding's attention into the following month, at which time Schröter began making his own observations. "The 3rd, 16th and 21st February saw no appearance of the night side," Harding wrote. "On the other hand, on the 28th at 6:12 pm, in late daylight and clear air, with the telescope mentioned above, the dark portion of the disc of Venus was seen without a doubt." This was, however, the very same date on which Schröter wrote that he "could not detect the slightest trace of the night side of Venus" due to a combination of twilight and moonlight interference at the time he looked.

Indeed, Harding quoted a time down to the minute for his claim of a night side sighting. On the last day of February 1806, the Sun set in Göttingen at 2 minutes before 6 o'clock in the evening or 14 minutes before the time at which Harding recounted that he saw "the dark portion of the disc of Venus … without a doubt." Venus then stood 19° above the west-southwest horizon at the time in very bright twilight, while the 77%-illuminated Moon hung far up in the southeastern sky. Given the brightness of the sky background against which Harding saw Venus, it seems implausible that what he described was a night side glow *brighter* than the sky around it. "This time, however," Harding wrote, "the brighter part did not appear that much larger than the dark part, undoubtedly because the observation on the 24th of January was made in complete darkness, but the present day was not yet ended, and the bright crescent was already much narrower than at that time." So what gave Harding the impression that he saw light brighter than the twilit sky?

A clue may be found by looking back into history. Andreas Mayer's 1759 account happened in the light of midday, and the few words he wrote for the record give little insight other than the impression that, somehow, something about the night side of Venus looked different than Mayer expected. The night side appearing *darker* than the surrounding daylight sky background is difficult to explain physically. But Harding noted a distinct hue to whatever he saw on the 28th, "a dull reddish-gray light, like the Moon in total eclipse." Other observers through history have noted discolorations of the unlit portion of the disc of Venus under bright conditions of the sky background, which have become a marker for suspected Ashen Light appearances in twilight and daylight.

Harding saw something again the following night, March 1, and that time he brought a friend:

> I immediately saw the night side of Venus with complete clarity during dusk with a 5-foot reflector, and this perception was absolutely confirmed in a 10-foot reflector. I never before saw this so clearly; the outlines were very sharply delineated, and were so marked against the darker blueness of the sky that the optician Gotthard also noticed it immediately at first sight.

His companion was Johann Zacharias Gotthard (1750–1813), a "helper" at the university observatory and maker of achromatic lenses for telescopes. Probably Harding brought him along that night as a kind of expert witness to ensure that his telescope was free of any obvious optical defects that might explain what he saw. By that point, Gotthard or no, Harding seems to have been personally satisfied by the reality of what he saw.

But what, for Harding, *was* that reality? Here we encounter one of the first instances where a firsthand observer of the Ashen Light attempted to work through the logic of his observation in print. "Regarding this unusual phenomenon the first question is probably whether it was real or a deception," Harding put it, an exercise of a reasonably dispassionate mind. After reading his account several times, I concluded that Harding was convinced that he detected something out of the ordinary when looking at Venus on certain nights and further that he *definitely* didn't see it on others. That much is clear. However, he wrestled with the nature of the thing that he saw and whether it was intrinsic to Venus or simply some kind of spurious phantom or illusion. If what he thought he saw were in fact illusory, it "would not explain its repeated observation, or the concordant statements of the completely unbiased optician Gotthard." What Harding didn't report was whether the two men saw the same phenomenon in an independent way. In other words, did one suggest the appearance of something unusual about Venus to the other before he had a chance to notice it on his own? We don't know.

To Harding, his own individual impression at the eyepiece was no more trustworthy than any other account from a lone observer. It only became real for him once he judged its independence sufficiently confirmed. On the first night of March 1806, he saw the faint glow of light again emanating from the night side of Venus, and this time he a credible associate at his side. Johann Gotthard's experience making optics, Harding reasoned, demanded some respect for what he plainly indicated that he saw, too.

Schröter's simultaneous report was the third leg of the stool of reason holding up the conclusion of the reality of the Ashen Light. Harding helped himself up onto that stool and proceeded to hypothesize about the cause of what the three men saw:

> When we perceive the dark part of the Moon's disc, which is averted from the sun, no one doubts that this phenomenon is due to the reflected light of the Earth; by itself, should the Earth also be able to send so much light to Venus that their whole night could be made visible from such a considerable distance? It is true that the light which Venus sends to our Earth is bright enough to make visible the shadow of the body which is to be seen, at a time when it is a crescent only

> one-eighth illuminated, as was the case on 24th January. As seen from Venus, the Earth was seven-eighths illuminated at the same time, and although the sunlight [at Earth] is only half as strong, it is yet still brighter than when we saw it, and the Earth's dense atmosphere should not alter it. Would not this phenomenon appear so often when light reflected from the illuminated part of Earth caused the visibility of the unilluminated part of Venus, when this planet comes into a similar position with respect to Earth and Sun? But even some of the most diligent observers of Venus have not seen it, for what the Count von Hahn once perceived does not seem to agree with the perceptions described above.[3]

Following this line of reasoning, Harding concluded that the Ashen Light must not be like the earthshine seen on the Moon. Rather, it must be due to the emission of light from something inherent to Venus itself. "I am," he wrote, "therefore inclined to consider this twilight in the night of Venus the effect of some phosphorescence, according to which this planet has to generate so much light from its surface that its nighttime hemisphere, even from such a distance, still can be seen."

He didn't take the further step of guessing the physical mechanism producing the light, for in Harding's time astronomers were largely ignorant as to what the visible face of Venus presented. There was not enough information to say, for instance, whether it was it a solid surface on which some observers thought they saw the shadows of mountain ranges cast in the slanted light of sunrise or sunset along the planet's terminator. Nor could they tell whether the planet was covered by a global ocean, merely reflecting sunlight as from a mirror, or if the globe of Venus was encircled by an atmosphere only whose cloud tops astronomers saw through their telescopes. Harding hedged his bet and merely offered that a "phosphorescence" of some kind must be responsible for the Ashen Light. If the process yielding it were intermittent or dependent on some other physical conditions, it would explain why the Light was seen with such infrequency. In his time, and with so little hard data available, it seemed like all options were on the table.

If Venus was anything like our own Earth, it might well manage to hold on to an atmosphere of some density, and for this reason Harding decided he couldn't rule out an atmospheric explanation. "According to the present state of our knowledge," he wrote:

> the impossibility of such a phosphorescence is probably not proven, since we know that the property of producing light is found everywhere in individual

[3] Harding quotes von Hahn in the *Berliner Astronomisches Jahrbuch für 1796*, p. 188.

> bodies of our Earth as well as in whole Earth itself. The bright twilight, which is often observed on moonless nights, and in which distant objects can be recognized quite distinctly, is probably nothing other than such a temporary phosphorescence of the Earth.

It's unclear exactly to which phenomenon Harding alludes here, but it qualitatively resembles airglow, a faint optical luminescence of the Earth's night sky caused by chemical processes in the upper atmosphere. Often called "nightglow" when seen specifically on the night side of Earth, airglow is distinguished from the aurorae—the Northern or Southern Lights—by a fundamentally different physical mechanism: while aurorae are generally limited to the regions near the Earth's poles and their visibility is strongly tied to the activity cycles of the Sun, some amount of airglow is present around the world on virtually any clear night. This accounts for why the night sky never appears a true, inky black, even in consideration of scattered starlight and highly diffused sunlight from the day side of Earth. Most people today are unfamiliar with the dim glow that Harding describes because its visibility requires a very naturally dark night sky, now unobtainable to much of humanity due to the influence of light pollution. But in Harding's time, and away from the major European capitals, airglow would have been noticeable to anyone who spent much time under the night sky.

Furthermore, Harding believed (often wrongly) that a number of astronomical objects radiated some kind of intrinsic light even in the absence of illumination by the Sun. For instance, he thought the "dull reddish light of the Moon during total eclipses" was caused by a lunar phosphorescence that had something to do with its surface characteristics and only became visible from Earth when the overwhelming glare of reflected sunlight was reduced, as when the Moon passed through the Earth's shadow during eclipses. Now, of course, we know what actually accounts for the reddish glow of the totally eclipsed Moon: sunlight streaming through the Earth's atmosphere and refracted around the edge of our planet, its blue rays scattered away by interactions with small particles in the air. It's the same reason why sunrises and sunsets are often reddish, especially on dusty days.

While it seems that Harding never settled on a firm hypothesis to explain what he saw toward Venus on a handful of nights in the late winter of 1806, his suspicion of a luminous atmosphere remained. If true, it would conveniently explain the intermittent nature of historical Ashen Light observations. "If this phenomenon is the effect of a phosphorescence," he reasoned, "a certain disposition of the atmosphere or of the surface of the planet is necessary in order to produce it, and the latter may fail for several years due to the coincidence of various circumstances."

Johann Wilhelm Pastorff (1822)

Harding's German contemporary, Johann Wilhelm Pastorff (1767–1838), was an astronomer best remembered for his many observations of the Sun between 1819 and 1833. Pastorff was of the landowning class, and he trained for a career in civil service but later retired in order to pursue his passion for astronomy full time. In staring at the Sun through telescopes for so many years, Pastorff created an invaluable record of sunspots that helped establish the reality of the 11-year solar activity cycle through which the numbers of sunspot groups wax and wane. In short, he was not a novice observer.

But Pastorff also reported seeing other phenomena that undercut the reliability of some of his observations. In 1819, for instance, he reported seeing the "Great Comet" of that year (C/1819 N1) transit across the face of the Sun. While a transit *did* take place as seen from Earth on June 26, 1819, the nucleus of a typical comet is far too small to have been detectable at its distance by a visual observer using a nineteenth-century telescope and its coma and tail too diffuse to have dimmed sunlight by any appreciable amount. Pastorff later claimed repeated observations of a small shadow transiting the Sun's disk over the course of several hours each time that was then widely believed to indicate the existence of a planet orbiting interior to Mercury. However, as we will find out in Chap. 8, no such planet ever existed.

Observations of planets were always a sideline to Pastorff's solar work, but he took a look from time to time and reported some rather odd phenomena. An 1822 paper he submitted to the *Berliner astronomisches Jahrbuch* was entitled "Further confirmation that Venus, Jupiter, and Saturn are surrounded by strikingly visible light spheres," which he began by immediately addressing his critics:

> With regard to the announcement of the existence of an important sphere of light revealed by me in the *Astronomische Jahrbuch für 1823*, pages 157–159 and 248, which is clearly visible on Venus, Jupiter, and Saturn even on moderately dark nights, I now endeavor to refute the doubts raised against it.

Even though what Pastorff saw at the eyepiece when he looked at these planets was certainly not real, his defense of the observations reveals something about the increasingly logical mind of nineteenth-century European astronomy. By this time, early journals were already publishing polite criticism and running dialogs of scientific disagreements. The notion of one's scientific reputation was at stake.

Pastorff wasn't very clear what he meant by "light spheres" around these planets, other than to note that they were "strikingly visible" to him. He describes what may be the first recorded instance in which an observer deliberately put the illuminated crescent of Venus just outside the field of view of the telescope in order to minimize the impact of scattered light in trying to detect faint emissions from the night side. Pastorff wrote that "the apparent light from the planetary discs is noticeable even upon rotation of the eyepiece" in order to suggest that what he saw was not some artifact of his telescope, implying that the phenomenon, while clear, is subject to vanishing under the slightest optical aberration.

"But this is even clearer and more conspicuous," he continued:

> as soon as the distinctly represented planet is then gradually removed from the field of the telescope, provided that it is made well and admirably, and by a master like Ramsden, Fraunhofer, Dollond, etc.,[4] and the edge of the planet can only touch the contour of the outside surface.[5] No observer can then fail to notice them. As I have already remarked, Venus is so clearly and brightly cut off from the heavenly space around it, as is the dark part of the moon, which is also visible to us in the light of the Earth in the field of the telescope.[6] The least distortion of the eyepieces, or the displacement of the lens, however, makes them disappear even if the planet is still clearly visible in the telescope.

Pastorff went to lengths to assure the reader that his telescopes were of such high quality that they "represent every object with extraordinary clarity."[7] He reports having taken care to remove as many sources of interference in the observation as possible, from avoiding conditions of "a roiling atmosphere" to blackening the inner surfaces of his eyepiece barrels in order to prevent stray

[4]This is a reference to some well-known designers of telescopes and eyepieces in the late eighteenth and early nineteenth centuries: Jesse Ramsden (English, 1735–1800); Joseph Ritter von Fraunhofer (Bavarian, 1787–1826); and Peter Dollond (English, 1731–1820).

[5]Pastorff here seems to refer to the practice of placing the bright crescent just beyond the curved edge of the field of view of the telescope, thereby suppressing the influence of light from the crescent that would otherwise scatter toward the night side.

[6]Pastorff here compares the quality of the light from the night side of Venus to the phenomenon of Earthshine.

[7]"The proof of the excellence of the 5 1/2-foot telescope is that stars of the first magnitude, such as Capella, appear completely rounded and close double stars appear completely separated."

internal reflections from their metal surfaces. “I took precautions to remove all of these influences,” Pastorff wrote, “and yet I still saw these light spheres.”

In keeping with the need to defend himself from critics, Pastorff thought about other circumstances of his observations that may have led him to incorrectly perceive the “light spheres.” He considered whether something about the night sky background might have led his eyes and brain astray. If the light source were not in the Earth’s atmosphere, what about something further afield? Besides diffuse starlight, the light of the Milky Way might interfere with the observation. But the apparent path followed by the Sun, Moon, and planets throughout our night sky doesn’t take them past the Milky Way for very long.

Another source of faint background light, however, *exactly* follows this path. Called the zodiacal light, it is the result of countless small particles in the plane of the Solar System, the minute leavings resulting from billions of years of comets shedding debris, and asteroids and other small Solar System objects grinding each other down to literal dust. Particles in solar orbits beyond our own scatter sunlight back in our direction, which we see from Earth as an ethereal, cone-shaped glow extending above the horizon before sunrise or after sunset. Pastorff considered this possible explanation for what he saw and dismissed it:

> A second argument against the existence of the discovered light spheres might be that Venus, Jupiter, and Saturn were at the time of observation seen against the background of the zodiacal light, and that by this immersion in it the light spheres seem to have originated, but this is by no means the case. For these planets have been observed by me most conspicuously with light spheres, even though they were very far separated from the zodiacal light, or the zodiacal light was not visible at all, and the air was very pure and clear.

What exactly *did* Pastorff see through his “excellent telescopes, which represent every object with extraordinary clarity”? His account is one of the first that mentions any sort of irregularity to the brightness of the Ashen Light, a patchiness in its appearance that led him to speculate on what caused it:

> I must remark, however, that as a proof of the great light-sphere existing around Venus, it may well be said that such a dazzling light could not possibly be seen at a time when the lit part of the planet was only very narrow and crescent-shaped, not half an inch long, if she did not in fact shine with her own light. Only this clearly showed the dark part of it in a soft gray light, so illuminated by the light sphere that I discovered, that in this dark part I discovered some large darker and lighter spots.

In effect, he argued for some local source of light on Venus that accounted for the inhomogeneous distribution of light across its night side. It was an effect, he argued, that "would not be possible without a light sphere, because reflected sunlight cannot do this, as any astronomer can see." But the exact substance of Pastorff's "light sphere" remained elusive, even as he searched for some natural analog on Earth. He leaned toward the combustion of some volatile substance in the Venusian atmosphere, a "ponderable air" that he thought might have something in common with the "luminiferous aether," a medium for carrying light waves from distant objects to the Earth thought at the time to permeate the cosmos. The aether was presumed to be some kind of incompressible fluid not subject to constraints imposed by objects that it surrounded.

A hypothetical phosphorescent substance, Pastorff thought, "differs from the light diffused in the universe; it is clear that the phosphorescent light must form a compressed light sphere through the centripetal force of the planetary bodies, which in any case will be more or less visible to us." The inconsistency of this process just might explain why the visibility of the Ashen Light was an unusual experience. He believed that the tails of comets shone via the same mechanism, arguing that the characteristics of the bright comets seen in 1807, 1811, and 1819[8] validated the notion of light spheres around other Solar System objects and implying that the same substance was present on (or near) both planets and comets.

Pastorff also seems to be the first observer of the Ashen Light who tried to estimate its surface brightness quantitatively. He assigned it a brightness of "100" on an arbitrary scale, which he defined as equal to the intensity of the earthshine on the Moon, and guessed the brightnesses of what he saw toward Jupiter and Saturn as 80 and 60, respectively. The Venus "light sphere" was well defined in contrast to the darker sky behind it, Pastorff wrote, and distinctly grayish-red in color.

Deciding whether any of this was real depends on one's judgment of the observer's reliability. Pastorff made many useful observations of the Sun during his astronomical career, none of which would now be considered especially controversial. But he characterized what he saw toward Venus in the same sense as what appeared to him around Mars, Jupiter, Saturn, and even the minor planet (1) Ceres; in the case of Jupiter, he reported that "every moon seems to be surrounded by its own light sphere." In the end, Pastorff didn't

[8]These are very probably comets C/1807 R1, C/1811 F1, and C/1819 N1 (Tralles), respectively. The "gleam of light" (*Lichtschimmer*) Pastorff describes is reminiscent of stellate apparent nuclei in the comae of many comets near where the tail is seen to begin.

report enough in the way of details to allow a firm assessment of the quality of his observations, and we are left to wonder what he actually saw.

Franz von Paula Gruithuisen (1825)

Elsewhere in Germany around the same time, one of Pastorff's fellow countrymen was making a name for himself in astronomy, although not exactly for the reasons he might have hoped. As an observer, Franz von Paula Gruithuisen's enthusiasm was outdone only by a vivid imagination that ultimately brought more ridicule than fame. The son of a Dutch father and a German mother, Gruithuisen (1774–1852; Fig. 4.3) began his career as an assistant surgeon in the Austrian army during the Austro-Turkish War. He later practiced in Munich at the court of Prince-Elector Karl Theodor of Bavaria (1724–1799) and for his service was sent to study medicine and philosophy in Landshut. He was a practicing physician for several years after graduation, inventing new techniques and devices for the treatment and removal of bladder stones. Since his student days, however, he harbored an obsession for astronomy that slowly turned into a career as he spent progressively more time at the telescope and less in the operating theater. After turning down academic appointments in Breslau and Freiburg, in 1826 he was made a professor of astronomy at Munich, where he spent the rest of his life.

Fig. 4.3 Franz von Paula Gruithuisen (1774–1852) seen in a contemporary portrait engraving. Wellcome Collection, licensed under CC-BY-4.0

Even before astronomy became his second career, Gruithuisen published prolifically. His tendency toward lengthy descriptions and attempts to interpret them usually led him to take things a little too far. For instance, based on his extensive telescopic observations of the Moon, Gruithuisen became convinced that it was an inhabited place and that evidence of structures built by an intelligent race was quite clear. In 1824 he published *Entdeckung vieler deutlichen Spuren der Mondbewohner, besonders eines collossalen Kunstgebäudes derselben* ("Discovery of many clear traces of the Moon dwellers, especially a colossal art building of the same"), in which he argued that apparent linear structures on the lunar surface could, in fact, have no other origin, rejecting the dominant belief at the time that the lunar features were probably of volcanic origin. His conviction in this was, he wrote, "strengthened more and more from day to day, that in the moon, as far as its surface presents us, there can be absolutely no trace of vulcanism, and that it would be in the highest degree absurd ... to declare the ring-hills [craters], which are twelve to fifty miles in diameter and above, to be crater-openings from which the moon should spit out all its contents." Instead, he was among the first to advocate for the proposition that lunar craters were caused by the impacts of small bodies like asteroids onto the Moon's surface, an idea that would not be proven conclusively until the middle of the twentieth century.

In the same way that his physical descriptions of lunar craters led him to support the ultimately successful impact theory, Gruithuisen's observations of linear—and seemingly, artificial—structures on the Moon brought him to the incorrect conclusion that they could only have been deliberately constructed. The most famous of the structures he saw on the Moon was a set of apparently raised lines north of the crater Schröter that he interpreted as a group of buildings separated by long, straight avenues (Fig. 4.4). He named it the "Wallwerk," interpreting a network of various, vaguely linear hills as a "large art temple in the shadow of Schröter." While conceding that some features in the vicinity were of natural origin:

> everything else in this figure is evidently the product of Selenite diligence. The ramparts here are usually over a mile wide, and while they may not exceed the highest buildings on the earth in height, they are usually several miles long. It is highly probable that they serve as dwellings of intelligent beings in the Moon.

Elsewhere on the Moon, Gruithuisen saw subtle differences in the surface hue as indication of vegetation and various climate zones. While wrong on all accounts, his prose often bordered on polemical and left the impression that he believed any idea that couldn't be proven *untrue* was as valid as something whose objective existence could be established beyond a doubt.

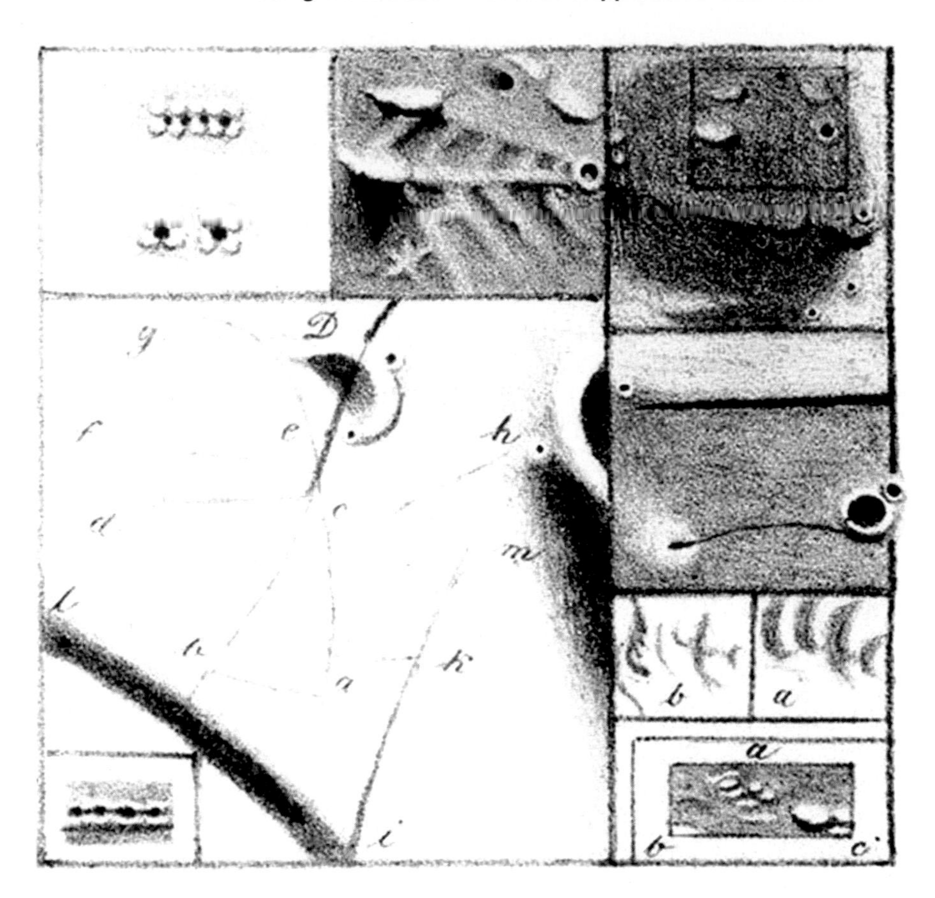

Fig. 4.4 Figures 2–9 from Franz von Paula Gruithuisen's *Entdeckung vieler deutlichen Spuren der Mondbewohner, besonders eines collossalen Kunstgebäudes derselben* (Nuremberg, 1824). These reproductions of Gruithuisen's sketches of various features seen in his telescopic observations of the Moon supposedly show evidence for artificial modification of the lunar landscape. The "Wallwerk," Gruithuisen's best hope for proving the existence of intelligent inhabitants, is seen with multiple radiating lines in the panel at top-center

In Gruithuisen's mind, many worlds were populated with intelligent beings. And to him, the Ashen Light was a clear indication of an intelligent race on Venus. He recalled his own observations nearly two decades later: "I have only seen it once on June 8, 1825, at 4 o'clock in the morning with the 30-inch Fraunhofer telescope at 60 and 150 times [magnifying power]. At first I thought that the twilight in front of the horns deceived me, but the phase was already too large."

That he only ever saw it once led him to believe that it must be some sort of extraordinary phenomenon. Besides his own observation, he seems to have been aware of only the accounts of Mayer (1759), Harding, and Schröter (1806). Noting that the time between these reports was something like a human lifespan, he reasoned that the appearance of the Ashen Light must reflect something happening on a comparable timescale:

> Between these epochs 76 Venus years or 47 Earth years have elapsed. While there is no justification to assume that this interval has a religious nature, it becomes more comprehensible if we assume that some Alexander or Napoleon then attained universal rule. If we assume that the ordinary life of a Venus inhabitant lasts 130 Venus years, which amounts to 80 terrestrial years, then the period of government of such a sole ruler may easily amount to 76 Venus years.

The logic led to a simple conclusion: "At the epochs when this faint light on Venus is visible, the inhabitants of the planet organize festivals and general illuminations, which are easier to arrange because the vegetation of Venus is incomparably more luxuriant than even the virgin forest of Brazil." In other words, the Venusian beings jubilantly burned down their forests upon the accession of a new monarch. Long live the king, indeed.

But in typical Gruithuisen fashion, he strained to make clear that he wasn't too attached to the idea even if it did seem to validate his beliefs about our planetary neighbor. "I have no intention to press this opinion and do not claim its credibility," he wrote, "even should it appeal to the reader's imagination; but if my hypothesis is correct, we at least receive direct testimony to the existence of inhabitants on Venus." Besides, he throws in for effect, global forest fires could be good for the Venusians. By thinning out the forests and creating newly arable lands to sustain a growing population, "large migrations of people would be prevented, and the consequent wars would be avoided and keep the tribes together." While Gruithuisen's interpretation of the Ashen Light raised more than a few eyebrows, it failed to gain adherents, and it was largely forgotten by mid-century.

The Later Nineteenth Century

Writing in 1855's *Astronomie Populaire*, François Arago summarized Ashen Light observations from Derham to Gruithuisen. In the pages of his book, he wondered aloud about its possible causes:

Can this rare and curious phenomenon not be explained by a certain Earthshine, similar to that of our Moon, and that would have its cause in the light reflected by the Earth or Mercury to Venus? Is not a more plausible explanation given by referring to the class of negative visibilities or by contrast? Can we attribute it to a kind of phosphorescence that sometimes develops in the material of which Venus itself is formed? Should we assume, finally, that the atmosphere of the planet is sometimes the location in its entirety [of the light emission], similar to those lights that, on Earth, are the Northern Lights?[9]

Yet still the qualitative descriptions of eyewitnesses failed to even hint at what was actually going on: "The observations so far," Arago judged, "have not proved specific enough for us to decide in favor of these assumptions, in preference to others."

As the subsequent decades rolled on, the number of reported Ashen Light sightings in the astronomy literature grew slowly but steadily. As an indication of how casual sightings of the Ashen Light had become among amateur astronomers of the Victorian era, consider a third-person account written by George Knott (1835–1894).[10] At the end of an 1862 article for the British journal *Monthly Notices of the Royal Astronomical Society* on his observations of variable stars, Knott made a throwaway comment about something his uncle up in Lancashire happened to mention to him, with emphasis in the original:

I take the opportunity of mentioning a recent observation of that curious phenomenon connected with the planet Venus, sometimes called the *phosphorescence of the dark side*.but On the evening of Jan. 14th, my uncle, Mr. Berry of Liverpool, was examining the planet with a small but fvery perfect Gregorian reflector of 4 inches aperture, mag. power 160. The wind was high, but the atmosphere was very clear, and in repeat intervals of quiet, when the cusps were sharply defined, the unilluminated part of the disk shone with a faint light similar in appearance to the *lumière cendrée* in the crescent Moon.[11]

To lend some credence to this secondhand tale, Knott appealed to his uncle's unfamiliarity with previous claims that might have influenced what he saw. "In proof of the independence of the observation I may say," Knott wrote, "that,

[9]Vol. 2., pp. 524–537.

[10]Knott was an English amateur astronomer from Cuckfield, Sussex, known largely for observations of double, multiple, and variable stars who "always showed the keenest in astronomical questions of all kinds." (*The Observatory*, Vol. 17, pp. 355–356, November 1894.)

[11]Knott here refers to Earthshine seen on our own Moon, using a contemporary French phrase for the phenomenon (literally, 'ashen light').

at the time, it had entirely escaped Mr. Berry's memory that the phenomenon had been remarked by previous observers."

Observing from Trincomalee in British Ceylon, now modern Sri Lanka, Percy Braybrooke Molesworth recorded an Ashen Light sighting in a moment when he was "not thinking of the phosphorescence of the dark side." Molesworth (1867–1908) was a Major in the corps of Royal Engineers and an amateur astronomer who made fine drawings of Mars and Jupiter during their oppositions in the years from 1903 to 1905. On the evening of October 6, 1898, Molesworth was observing Venus with a 12.5-inch (32-cm) reflecting telescope when he:

> suddenly noticed that the unlighted part of the disc was visible near the S cusp and also for some distance from the centre of the terminator, browner but slightly lighter than the sky. I could not trace the outline of the unilluminated circle except for a short distance from the S cusp. Its diameter appeared considerably less than that of the unilluminated part.[12]

Although he searched for the effect on the following nights, he never again reported it.

The Reverend Thomas William Webb (1807–1885), an Anglican vicar who pursued astronomy on the side as a hobby, cataloged some then-recent claims of Ashen Light observations in *Celestial Objects For Common Telescopes* (1873), including that of Knott's "Mr. Berry":

> Guthrie[13] and others noticed it a few years ago, with small reflectors, in Scotland; Purchas,[14] at Ross, in England; De-Vico[15] and Palomba,[16] many times, in Italy.

[12]Quoted by R. Baum, 2007, Insights into enthusiasm: The 1897–1898 Venus notebooks of P. B. Molesworth, *JBAA*, 117(1), 18.

[13]Little is known of "Mr. Guthrie, a gentleman residing near Bervie" in Scotland (1854, *Monthly Notices of the Royal Astronomical Society*, 14, 169). Guthrie reported being "unexpectedly surprised by observing an annular fringe of light surrounding the dark side of the disk, and completing the circle which was partially formed by the outer margin of the crescent." Richard Baum put this observation in around 1842, and identified it as indicating not the Ashen Light, but rather the elongation of the crescent cusps due to the forward scattering of sunlight in the atmosphere of Venus when seen very close to inferior conjunction. For an example of this phenomenon, see the series of photographs in Fig. 3.4.

[14]William Henry Purchas (1823–1903) was better known as an English botanist than an astronomer. Augustin Ley memorialized Purchas with a short biography in September 1905, *Transactions of the Woolhope Naturalists' Field Club*, Hereford: Jakeman and Carver, 341–344.

[15]Fr. Francesco de Vico (1805–1848) was an Italian astronomer and Jesuit priest who discovered a number of comets in the 1840s.

[16]Clemente Palomba (1819–1891) was an Associate Astronomer at the Observatory of the Gregorian University in the Collegio Romano at Rome. Palomba carried out an extensive Venus observing program between 1839 and 1841; from his measurements, totaling more than 11,000 in 1839 alone, Francesco de

Berry at Liverpool saw it in 1862 with no previous recollection of its visibility: it was remarked in that year, and especially in 1863, by many observers in England, and by one in 1865, as well as by four at Leipzig with the 8 1/2 in. achromatic. … The dark side is frequently too small in proportion, like that of the crescent Moon to the naked eye; and from the same cause,—the irradiation of the luminous part; it is sometimes described as grey, sometimes reddish. It would be well worth looking for when the crescent is narrow, but Venus should have high N. latitude to clear the vapours of the horizon.

The Ashen Light was eventually so widely recognized that it became the stuff of standard astronomy textbook fare. Yet in *Astronomie Populaire*, Arago cautioned about loose interpretations of what the reports implied. "There are not, in all these observations," he opined, "the elements necessary to decide to what must be attributed the unusual appearance of the portion of Venus not illuminated by the sun." The memory of Gruithuisen's explanation was still fresh, and Arago seemed skeptical in appealing to the plausibility of various explanations.

Another giant of nineteenth-century European astronomy and Arago's fellow Frenchman, Camille Flammarion (1842–1925) had some things to add in his own coincidentally titled *Astronomie Populaire* (1880). While called *lumière cendrée*, evoking the phenomenon of Earthshine:

it has no satellite to produce it. It seems to me that this visibility, rather subjective than objective, arises from clouds on the planet, which whiten its disc and vaguely reflect the stellar light scattered through space. The eye instinctively continues the outline of the crescent, and imagines, rather than sees, the rest. Besides, the aurora borealis may sometimes light up during the night the glowing sky of Venus, and its clouds may emit a certain phosphorescence, as we occasionally notice on the earth in the evenings of April and May.

Flammarion here pokes around the edges of the phenomenon for signs of its existence independent of the mind and eye of the observer, which maybe "imagines, rather than sees" light that is not there. His view began to find adherents in the last quarter of the nineteenth century.

Still, it seemed impossible to toss the Ashen Light aside as some sort of fictitious specter existing only in the imaginations of those who claimed to see it: there were just too many reliable observers who were certain enough of what they saw to publish their reports. Reviewing known appearances dating

---

Vico claimed a rotation period of about 23 hours and 20 minutes. See S. C. Chandler, 1897, Historical note on the rotation of Venus and Mercury, *Popular Astronomy*, 4, 393–397.

back to Riccioli in *A Popular History of Astronomy During the Nineteenth Century* (1893), Agnes Mary Clerke (1842–1907), renowned for her syntheses of contemporary astronomy research results as well as her writings on the subject's history, noted "the reality of the appearance" that "has since been authenticated by numerous and trustworthy observations." While discussing various explanations for the effect that were then in vogue, she appealed to a lack of definitive data and refused to advocate strongly for one mechanism or another. But she offered one of the first speculations on the sensory experience of the Ashen Light for a hypothetical observer on the night side of Venus: "Whatever the origin of the phenomenon, it may serve, on a night-enwrapt hemisphere, to dissipate some of the thick darkness otherwise encroached upon only by 'the pale light of stars.'"

Adalbert Safarik's 1873 collection of recorded Ashen Light accounts up to his time remained for years the go-to review of the affair. Counting mentions by 22 observers going back to Riccioli, Safarik drew some inferences from the limited number of reports then available, searching methodically for something among them that would point the way toward a solution. Among observations then extant, he noted that 13, or about 60% of the total, were made during twilight, while half saw the phenomenon in daylight as a discoloration of the unlit portion of the disc of Venus against the sky. He further broke down these observations according to the number of times a given observer claimed to have seen the Ashen Light. For twilight appearances, he found "once by 4, many times by 9," while for daylight observations the record was "once by 6, many times by 5." He continued:

> 4 observers saw a faint line of light encircling the dark disk, 19 of them saw the disk itself. Of the 22 cases reported, 12 have been observed during the last 11 years, say one per year; and I am disposed to think that the phenomenon is a normal one, and that with sufficient optical power and attention under a favourable sky it is to be seen at every inferior conjunction, though I would by no means advance that it is constantly visible, which would be a statement directly opposed to facts.

Astronomers were not unreasonably frustrated. Astrophysics was then opening up the universe to progressively better human understanding, at least insofar as the tools of classical physics could provide. That there were still relatively few published reports made the picture of the Ashen Light ripe for speculation; still, astronomers who paused for some thought on the physics of the situation saw glimmers of hope.

"If the phenomenon be a real one," wrote S. Maitland Baird Gemmill in 1895, "it would appear as if we were thrown back upon the idea of an intermittent cause occurring in the higher regions of the planet's atmosphere."[17] In other words, then-current knowledge about geophysical processes might be brought to bear on the problem. Gemmill's logic led him to believe that the origin of the Ashen Light must be found in the upper atmosphere of Venus and that an earthly analog fit the bill: "If the processes going on in Venus resemble terrestrial processes, then it is fair enough to infer that Venus may have its cycles of magnetic phenomena, and that these may be on a specially magnificent scale with corresponding auroral manifestations." After all, given the size of Venus and its proximity to other small, rocky worlds like Mercury and Earth, why *shouldn't* Venus have its own permanent magnetic field driving aurorae?

But Ashen Light observations to that point in history didn't seem to correlate strongly with either the phase of the solar magnetic activity cycle or the frequency of strong auroral displays on Earth. If the same underlying physics were involved, there must be something peculiar about the situation at Venus that didn't work the same way as it did on our own planet. Still, it wasn't an explanation on which Gemmill was willing to fully hang his hat: "I suspect, however, that the recorded observations embrace more than one kind of phenomenon."

Could modern scientific instruments solve the mystery? "Zöllner[18] has expressed his conviction that under spectroscopic examination, the ash-colored secondary light of Venus will be found to present bright lines," the editors of *Nature* offered in June 1876, "and it may be hoped that opportunities for such observations may occur during the present summer." Spectroscopy was by this time a workhorse of the physics laboratory, but it was still new to astronomy owing to inefficient optical designs and exceptionally "slow" photographic emulsions. The spectroscope uses refracting or diffracting optics, such as prisms and gratings, to disperse white light into its constituent colors. Hiding in among the various wavelengths of light arriving at the Earth from celestial sources are telltale signs of all kinds of astrophysical processes, as light of particular colors is added to or subtracted from the train of light waves along

[17] 1895, *JBAA*, 5, 412–414. Little about this Glaswegian amateur astronomer (*c.* 1860–1911) remains in the historical record. He was a Fellow of the Royal Astronomical Society and a founding member of the British Astronomical Association. As an observer, he made contributions to the literature on "the aurora, the zodiacal light, the visibility of the dark side of Venus, [and] the Milky Way." (1911, *JBAA*, 21, 197).

[18] Johann Karl Friedrich Zöllner (1834–1882) was a German astronomer known for his studies of optical illusions who also dabbled on the side in parapsychology.

the way from source to detector. And it might indicate the stuff of which the Ashen Light, if real, was made.

Already, spectroscopy of Venus had revealed a puzzle: its visible surface appeared to be a perfect mirror of sunlight, making no modifications to light on its way in or out of the atmosphere. Referring to the chemical fingerprints in the atmosphere of the Sun measured precisely through Earthbound spectroscopes, Julius Scheiner (1858–1913), an expert in astronomical spectroscopy at the University of Berlin, noted in 1894 that "Vogel,[19] Secchi,[20] Huggins[21] and others have compared a great number of the Fraunhofer lines on the spectra of Venus and the Sun and have been unable to detect the slightest difference."[22] However, the existence of known absorption features attributable to Earth's atmosphere in the spectra of astronomical sources made it seem reasonable that something like the same should be happening at Venus:

> and hence the nature of the two atmospheres must be similar. The faintness of the atmospheric lines of Venus indicates that the atmosphere is very thin, or else that the sunlight can penetrate only a short distance into it, being thus reflected from its upper strata. The latter explanation agrees well with other astronomical observations which show a thick envelope of clouds prevents a view of the true surface of the planet.

But spectroscopists kept searching, and by the early 1930s they identified absorption due to carbon dioxide—the atmosphere's chief constituent—in high-resolution spectra of the daytime side of Venus. There was hope that sufficiently sensitive photographic plates might eventually pick up traces of light from the night side that could be matched up to laboratory references

[19] Hermann Carl Vogel (1841–1907) was a German astrophysicist who made the extensive use of spectroscopy in astronomy, analyzing the atmospheres of stars and planets and measuring a precise rotation period for the Sun on the basis of the velocity shift of spectral lines in its atmosphere. Vogel also reported his own observation of the Ashen Light in November 1871 and had an idea about what explained it; see Chap. 5.

[20] Angelo Secchi, previously encountered in Chap. 1, is mainly remembered for devising the first system of stellar classification according to features seen in the spectra of stars.

[21] Sir William Huggins (1824–1910) was, with Vogel and Secchi, among the founders of astronomical spectroscopy who worked alongside his wife, Margaret Lindsay (Murray) Huggins (1848–1915). The two published *An Atlas of Representative Stellar Spectra* (1899), which remained a standard text for several decades.

[22] 1894, *Die Spectralanalyse der Gestirne* ("A treatise on astronomical spectroscopy"), Boston: E. B. Frost, Ginn and Co. The "Fraunhofer lines" are a set of strong absorption features in the solar spectrum first identified by the German physicist Joseph von Fraunhofer (1787–1826), who although not the first person to see the lines was the first to systematically study them and measure their wavelengths.

in hopes of identifying the "carrier" of the Ashen Light. If it turned out to be something like atomic oxygen or nitrogen, perhaps the Ashen Light was just the sort of run-of-the-mill aurora some observers had suggested.

By the end of the nineteenth century, human understanding of the planets had reached something of an impasse: bigger telescopes were not necessarily yielding better observations, while planetary images on photographic plates remained indistinct and offered no significant time resolution. Meaningful progress needed the robot eyes of spacecraft making close approaches, but such visits were still decades away in the future. The sum of knowledge about the planets gathered since antiquity was that they were worlds unto themselves and not merely the "wanderers" of the ancients. Some were attended by their own systems of satellites like Solar Systems in miniature. Some had rings and atmospheres; others, it was speculated, might harbor life. But they were all equally inscrutable beyond what little the telescope could see through the turbulence of Earth's atmosphere.

Perhaps the least was known with any certainty about Venus, but something Earthbound observers saw through the eyepiece at least seemed unique to our nearest planetary neighbor: the Ashen Light. "May I suggest that we are badly in need of a plausible theory of the cause of this phenomenon?" asked Francis James Wardale (1856–1942) in an 1895 letter to the *Journal of the British Astronomical Association*.[23] "We have tenable hypotheses, probable theories, or at least plausible suggestions about most astronomical mysteries, but I do not think anyone has yet proposed a solution of this difficulty having any approach to probability." And as the new century dawned, it appeared as though astronomers were further than ever from definitively explaining it.

[23] *JBAA*, 6, 35. Born in Wiltshire, Wardale's varied interests included "cricket, astronomy, meteorology, and field botany." (1943, *Monthly Notices of the Royal Astronomical Society*, 103, 73) He was known particularly for his studies of double stars and the planet Mars. Wardale was baffled by the unpredictability of Ashen Light sightings, reasoning that "surely in the twilight it ought in such a case to be an easy object to everyone in reasonably good air. No theory of the cause can be regarded as complete which fails to account for this peculiarity."

5

What Is the Light? Historical Explanations of the Ashen Light

Before diving into the possible reasons for why something on the night side of Venus may be glowing, it's worth reviewing the evidence from the historical record to date. There are eight observed facts about the Ashen Light:

One, it's clear that the Ashen Light is not something that is "on" all the time; rather, its appearances are episodic and seemingly unpredictable.

Two, great names in the history of observational astronomy claimed to have seen it, while others stated clearly that they *never* saw it.

Three, whether or not observers report seeing it doesn't seem to depend on either their locations or which telescopes they use. Some have noted it more or less simultaneously through different instruments, while in other instances it was seen on the same date by several unconnected observers in different places.

Four, the Ashen Light is most often reported in the 4 weeks immediately before inferior conjunction, but that fact might simply be the result of a selection bias that favors the convenience of evenings over mornings for Venus observations.

Five, there is a weak correlation between Ashen Lighting reports and the phase of the solar cycle in which the likelihood of sightings is somewhat higher around solar maximum than solar minimum.

J. C. Barentine, *Mystery of the Ashen Light of Venus*, Astronomers' Universe,
https://doi.org/10.1007/978-3-030-72715-4_5

Six, eyewitnesses describe the apparent color of the Ashen Light inconsistently, ranging from ashen gray to vaguely greenish to dull red to bright copper. In few instances do observers report no sensation of color at all.

Seven, there are reports of discolorations of the disc of Venus, opposite the bright crescent, during daytime observations around the time of inferior conjunction. But it's unclear that this phenomenon is of the same nature as situations where the night side of the planet is brighter overall than the (night) sky around it.

And eight, no one has yet obtained any photograph, image, or spectrogram that objectively, unambiguously, and convincingly records the existence of the Ashen Light.

The correct explanation for the Ashen Light must account for all eight of these facts. With that in mind, let's consider some of the possibilities.

Reflected Light

Among the earliest interpretations of the Ashen Light is one drawn from an experience familiar to anyone who has seen the Earth's Moon in our skies when it's a slim crescent on either side of the Sun shortly before sunrise or after sunset. It is the distinct appearance of a faint glow from the "night" side of the Moon, which is not illuminated directly by the Sun (Fig. 5.1). On a solid body with only the faintest trace of an atmosphere like the Moon, we would expect the night side to be almost completely black since the landscape would only be illuminated very weakly by the integrated light of stars, the Milky Way, and any planets other than Earth that happened to be above the local horizon. Instead, we see enough light that the night side of the Moon is clearly discernible by the unaided eye against the darker background of Earth's night sky.

Although the phenomenon was known at least from antiquity, it appears that the first person to correctly explain it was the Italian polymath Leonardo da Vinci (1452–1519), who realized that the light was, in fact, sunlight reflected from the Earth seen in reflection from the Moon's surface. This effect is most distinct when the Moon is "inferior" to the Earth relative to the Sun, meaning that the Earth is further away from the Sun than the Moon and its surface lit more fully. Given that the Earth subtends a fairly large angular size in the Moon's night sky and our planet is often covered with highly reflective clouds, it's easy to see why this explanation makes sense.

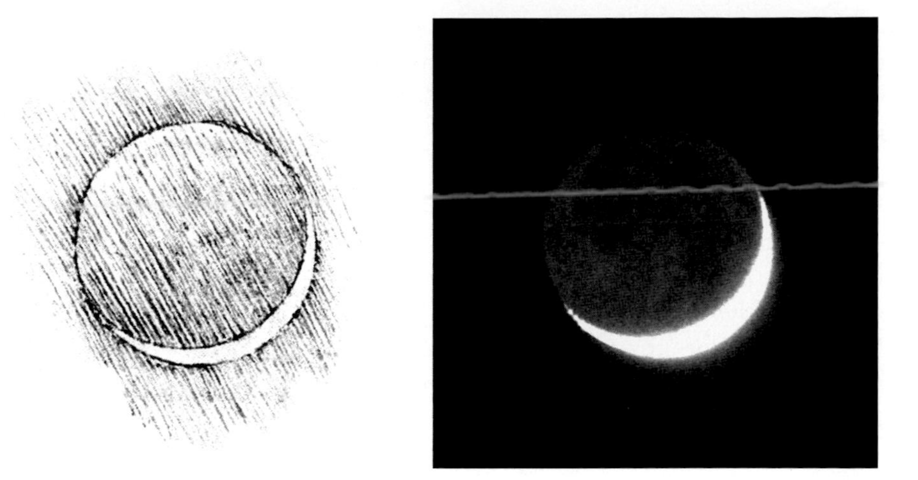

Fig. 5.1 *Left*: Leonardo da Vinci's impression of the earthshine as recorded on Folio 2 (recto) of the *Codex Leicester* (*c.* 1506–1513). *Right*: A photograph of the Moon taken by the author at a similar phase

The cause of the "earthshine" was understood before the invention of the telescope, and da Vinci's explanation would have been known to early telescopic observers of Venus. The earliest accounts likening the Ashen Light to earthshine were not unreasonable in the sense that it's the simplest explanation given the facts: faint light seen to be emanating from the night side of Venus is probably the reflection of light from some astronomical body other than the Sun. However, there's one important problem, as noted by Patrick Moore in *The Planet Venus* (1961): "If the Ashen Light really exists, it is not easy to account for, in view of the fact that Venus is moonless." While multiple claims were made for telescopic sightings of a moon of Venus between 1645 and 1768, no modern scientific evidence vouches for its existence. In any case, natural satellites of Venus invisible to terrestrial telescopes would also be far too small to account for an illumination effect bright enough to be seen from Earth.

In 1859, a German schoolteacher and inventor named Joseph Rheinauer proposed a modification of this idea in which the Earth, as the next large Solar System body to Venus, was the source of the Ashen Light, noting that the illumination of Venus by the Earth near inferior conjunction is roughly 480 times fainter than the illumination of the Earth by the full Moon.[1] This

[1]1859, *Die Erleuchtung des Planeten Venus durch die Erde*, Freiburg: Wagner. The figure of 480 times is quoted in Rheinauer, 1862, *Grundzüge der Photometrie*, Halle: H. W. Schmidt, 56; see also C. V. Zenger, 1863, On the visibility of the dark side of Venus, *Monthly Notices of the Royal Astronomical Society*, 43, 331. Of Rheinauer himself, very little is preserved in the historical record. He lived in Offenburg, then

scenario might explain why historical observations of the Ashen Light tended to cluster in the few weeks immediately before and after inferior conjunction, during which time Venus approached Earth more closely than at any other point in its orbit.

Rheinauer appealed to the simplicity of his idea as the reason it was likely correct: "If, by the way, we have succeeded in showing that the explanation of the phenomenon, according to the observations so far, is compatible with the assumption of Earthshine, our purpose is fully achieved by keeping to the principle—'the causes of natural things are not to be further admitted than those which alone are sufficient to explain their phenomena.'"

Citing the fact that observers only seemed to report the Ashen Light during times immediately adjacent to inferior conjunction, Johann Hieronymus Schröter wrote in 1806 that "if reflected earthly light did not cause the visibility of the dark side of Venus, would this phenomenon occur just as often as this planet comes into a similar position against the Earth and the Sun?" But Patrick Moore pointed out that "this explanation is clearly inadequate, since earthlight on Venus would be too weak to cause any perceptible glow." Ellen Mary Clerke noted in 1893 that "'full earthlight' on Venus, at its nearest, has little more than $\frac{1}{12000}$ its intensity on the moon," in agreement with Moore that "we see at once that the explanation is inadequate."

Even Adalbert Safarik dismissed the earthshine hypothesis, noting that the Earth, "as seen from Venus, far exceeds the greatest brightness of Venus as seen by us." While Venus on average reflects much more incident sunlight back into space than does the Earth, it is only a few percent illuminated by the Sun as seen from Earth when it makes its closest approaches to our planet. Safarik further quotes Rheinauer's calculation that the flux of the Ashen Light, "if resulting from this cause, should equal a star of the fourteenth magnitude.[2] That this explanation is insufficient is so clear as to need no further proof." In any case, sightings of the Ashen Light have been reported at times when Venus

in the Grand Duchy of Baden and now a town in Baden-Württemberg, Germany, and was active in the mid-nineteenth century. He taught physics at the Offenburg Gymnasium and dabbled in astronomy.

[2]Astronomical magnitudes assign a number to objects like stars according to their brightness. The magnitude system is unfamiliar to most people in part because it is logarithmic in nature: one integer step in the system indicates a difference in brightness of one power of the base number 2.512, chosen such that the system characterizes the human perception of the brightnesses of stars well. The brightest star in the night sky, Sirius, has an apparent visual magnitude of about −1.5; Rheinauer's comparison here, at magnitude +14, is some 1.6 million times fainter to the eye.

was relatively far from Earth,[3] and it's not consistently noticeable when Venus is near our planet.

Bill Napier, an astronomer now at the University of Buckingham in England, ran a more detailed calculation in 1971. He computed that the Earth as seen from the cloud tops of Venus near inferior conjunction has a visual surface brightness, or luminance, of about 1.6×10^{-4} candela per square meter (cd/m^2), noting that an assumed night sky brightness of 3.2×10^{-3} cd/m^2 yields a telescopic detection probability of 50%.[4] But since the incident brightness of Earth light at Venus is at best about 20 times below the threshold at which an observer with a small telescope is likely to detect it, the earthshine hypothesis can be safely discarded.

Aurorae

Early thinking about the source of the Ashen Light linked observations to terrestrial phenomena, which is a thoroughly understandable approach: while we might incompletely understand Earth, we know a lot about it from our up close and personal experiences gleaned while living on our world. The Earth is a decidedly much more accessible laboratory than the atmospheres and surfaces of other worlds.

Just as the reflected light hypothesis was born from an appeal to a terrestrial analog in the form of earthshine, some scientists looked to the Earth's atmosphere to explain where faint, diffuse (and ephemeral) light might originate. In the higher latitudes of our planet, that light comes from the aurorae, commonly called the northern or southern lights. "This was partly Schröter's idea," Adalbert Safarik wrote in 1874.

[3]Both Henry McEwen and R. E. Pressman reported observations of the Ashen Light in December 1939, some 200 days before the following inferior conjunction of June 26, 1940, and 380 days after the preceding inferior conjunction of November 20, 1938.

[4]Napier assumed the detection thresholds based on the pioneering laboratory studies of Blackwell (1946, Contrast Thresholds of the Human Eye, *Journal of the Optical Society of America* 36(11), 624–643) and Middleton (1957, "Vision through the Atmosphere" in *Geophysik II / Geophysics II. Handbuch der Physik / Encyclopedia of Physics*, Vol. 10 / 48. Springer: Berlin, Heidelberg). For reference, a natural, unpolluted night sky on Earth has an average luminance of about 1.71×10^{-4} cd/m^2, or 22 magnitudes per square arcsecond.

> It is supported by a most extraordinary observation of Mädler,[5] who, during the whole evening of April 7, 1833, saw Venus surrounded by long bright immovable rays. Professor Zöllner, of Leipzic, strongly advocates this idea and trusts that the spectroscope will reveal bright lines in the gray light of the unilluminated hemisphere of Venus.

Described as early as the fourth-century BCE and believed in antiquity to be an early indication of the coming day (and, hence, derived from a Latin term meaning "dawn" or "morning light"), a correct scientific explanation for the aurora wasn't articulated until the turn of the twentieth century. It relies on something both the Sun and Earth have in common: an intrinsic magnetic field.

Magnetic fields are most familiar to us as a navigational aid: a compass needle aligns north-south no matter how the compass is oriented with respect to azimuthal direction and thus shows us the way toward either of the Earth's poles even in unfamiliar territory. It does this because the needle is itself magnetized. Magnetic fields are so-called 'vector' fields that carry a sense of innate direction; the magnetized compass needle feels a force when misaligned with the local direction of the Earth's field and rotates around a pivot point until the force is at a minimum.

The needle keeps its magnetism because of some quirks of quantum mechanics having to do with the characteristics of the electron. But we know that's *not* what's happening with the Earth; in other words, our planet isn't a scaled-up version of a refrigerator magnet. We also know some basic properties of the Earth's magnetic field: it's hundreds of times weaker than a refrigerator magnet; its direction is currently out of alignment with the spin axis of the Earth by about 10°;[6] and its intensity and local direction vary according to one's location on the planet. And for reasons we don't completely understand,

---

[5]Johann Heinrich von Mädler (1794–1874) was a German astronomer who, together with Wilhelm Wolff Beer (1797–1850), published *Mappa Selenographica*, the first scientifically accurate map of the Moon, in four volumes between 1834 and 1836. For a more complete account of what von Mädler reported in 1833, see R. Baum, 1988, Historical Note: The Maedler Phenomenon, *Journal of the British Astronomical Association*, 98(6), 293.

[6]The north magnetic pole is now moving nearly 50 km a year relative to the Earth's rotational north pole; since its formal identification in 1831, the north magnetic pole has accumulated a displacement of about 2,000 km from the Boothia Peninsula in the Canadian far north to a position high in the Arctic Sea. The magnetic pole's rapid motion and a decades-long observed decrease of the intensity of the Earth's magnetic field have led some to speculate that the planet is about to undergo a reversal of its intrinsic magnetic field direction. These geomagnetic reversals have occurred randomly some 183 times over the last 83 million years; see, e.g., S. C. Cande and D. V. Kent, 1995, Revised calibration of the geomagnetic polarity timescale for the late Cretaceous and Cenozoic, *Journal of Geophysical Research*, 100, 6093–6095.

it sometimes weakens to almost nothing before being reborn, but oriented in a different direction.

The reason for all this has to do with electric currents. Run electricity through a long wire, for example, and the needle of a nearby compass will deflect slightly from its alignment with the Earth's magnetic field; shut the current off, and watch the needle move back to its starting position. Or, run a bar magnet through the turns of a coil of wire and a current will start to flow in the wire—but it disappears the instant that the magnet stops moving. The realization that electricity, magnetism, and light were all manifestations of the same underlying natural phenomenon was a defining moment of nineteenth-century physics, and their mathematical unification helped explain all of these key observations. It also showed why the Earth itself is the source of a magnetic field.

Through a process called differentiation, over billions of years heavy materials like iron have sunk to the core of our planet while light substances like silicate rocks floated up to the surface. As the iron was on its way down, it brought along a number of naturally radioactive elements whose slow decay deep within our planet has released a lot of heat over the Earth's history. High temperatures and pressures in the outer core keep these materials in a liquid state, and the heat, along with the planet's rotation, forces the molten materials to circulate. The molten outer core is also electrically conductive, and swirling motions within the material cause currents to flow. Electric currents yield magnetic fields oriented perpendicularly to the direction of those currents; since the direction of the current circulation is in roughly the direction of the core's rotation, the magnetic field emerges more or less parallel to the Earth's rotation axis. But the field wants to form closed loops, so it connects with the lines emerging from the opposite magnetic pole.

Our planetary magnetic field is maintained through a feedback loop involving all elements of this unified electromagnetic field: loops of current in and near the core generate a magnetic field, which as it changes sets up an electric field, which in turn exerts a force on the charges that are flowing in the currents. As long as the core remains at least partially molten, and as long as the Earth remains spinning on its rotational axis, the magnetic field persists through this self-sustaining "dynamo" effect.

Similarly, the Sun is made of a hot, convecting, and electrically conducting fluid, and it exhibits a persistent magnetic field through its own dynamo. Its strength rises and falls over a period of about 11 years, and at the start of every other cycle its direction reverses. Unlike the mostly solid Earth, the Sun rotates like a fluid throughout its interior so that its rotation period depends on latitude. This tends to tangle up the magnetic field lines as they

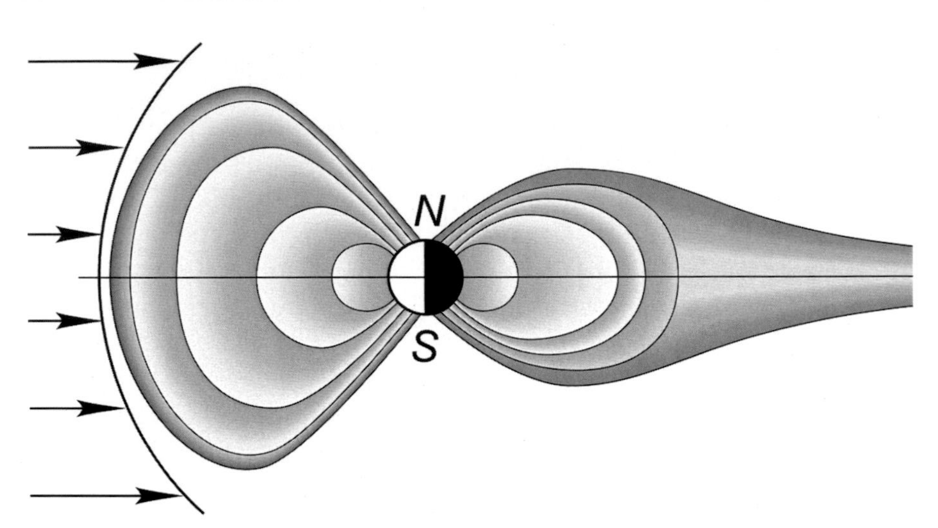

Fig. 5.2 Magnetic field lines generated in the core of the Earth (gray contours) are shaped by the magnetic pressure of the solar wind, whose inbound direction is shown by the arrows at the left. "N" and "S" indicate the north and south rotational poles of the Earth, respectively; the magnetic poles are oriented in the opposite sense. Graphic by Alec Baravik

try to force their way out of the Sun, winding them up like so many twisted rubber bands. Occasionally these magnetic "rubber bands" reach the point of breaking, flinging hot plasma out into space. Gradually the Sun loses mass to the escape of electrically charged particles, which are pushed outward into the Solar System as a solar "wind." In the absence of pressure exerted on the field by the solar wind, the terrestrial magnetic field lines would form neat, nearly circular loops around the Earth. But the bulk motions of these charged particles—the breezes of the solar wind, if you will—buffet and shape the magnetic environment around our own planet, drawing the Earth's magnetic field lines out into a long tail forever facing away from our star (Fig. 5.2).

Some of those energetic charged particles from the Sun, mostly protons, are carried by the solar wind into the Earth's magnetosphere, the region around the planet where our own magnetic field dominates over the influence of the interplanetary magnetic field. The rules of electromagnetism again take over: as the particles approach the Earth, the magnetic field exerts a force on them that is at all points perpendicular to the direction of both their travel and the field. The result is a spiral motion by which a given particle travels down a given field line, inevitably drawn closer to the Earth's surface. This would be all fine and well if our planet was devoid of an atmosphere; luckily for us, it's not. Near the top of the atmosphere, the molecules are pulled apart by collisions and exposure to ultraviolet light from the Sun; the high temperatures help strip electrons from their constituent atoms, leading to a soup of particles, both charged and neutral.

The solar wind particles move through this medium effortlessly at first, but the real struggle begins further down where the density of gas increases. Occasionally a solar particle slams into a free atom of, say, oxygen or nitrogen floating near the edge of space at altitudes around 100 km, imparting its energy to that atom. A series of quantum mechanical rules then kicks in, allowing the atom to exist in some energetic states but not others. While bearing the brunt of the particle impacts, an atom can find itself in an "excited" state in which it briefly holds on to some of the mechanical energy in the solar wind. Nature, however, doesn't like this condition very much, and it wants systems like atoms to settle into the lowest stable energy state possible. Nature also likes energy to be conserved, that is to say, the total available amount of energy in the universe is fixed; therefore, energy is allowed only to change forms. Excited atoms give their energy back to the cosmos in the form of light—electricity and magnetism, unified—before dropping again into their lowest-energy or 'ground' state. When enough of these collisions are happening with sufficient frequency, the energy from all of the atoms relaxing to their ground states is emitted as light intense enough to be seen to the unaided eye as shimmering curtains of green, red, and purple across far northern or southern night skies. The colors are a clue to the kind of atom involved, a kind of fingerprint assigned to each allowable energy transition within each kind of atom.

The aurora rises and falls in intensity with the strength of both magnetic activity on the Sun and strong electric currents in our upper atmosphere. While at least some weak amount of auroral activity is visible near both magnetic poles of Earth on any given night, there are instances around the peak of the solar activity cycle when it may become visible from the tropics. Within any night, the brightness of auroral displays waxes and wanes on timescales of seconds to hours, growing brighter or fainter depending on the orientation of the local interplanetary magnetic field and the flux of particles coming from the Sun.

As a potential mechanism to drive the Ashen Light, the aurora has a distinct advantage: it's a thoroughly understood phenomenon with a clear analog here on our own planet. The Belgian physicist Pierre-Jacques-Frédéric de Heen (1851–1915) worked through the logic in 1872 after considering several potential explanations for the Ashen Light including zodiacal light, phosphorescence of the Venus atmosphere or surface, backscattering of light from diffraction around the shadow cone trailing behind the planet ("which can form a curtain of light projecting on the dark part of the planet"), and something like the terrestrial aurora. de Heen settled on the latter explanation, he wrote, because

> it was found that terrestrial magnetism acquires its maximum intensity when the Earth is closest to the Sun. It is therefore easy to conclude from this that the intensity of the magnetic currents on the surface of Venus, which is much harried by the Sun, is also much greater. And one can not deny the intimate relationship between the intensity of magnetic currents and the intensity of the aurora borealis.

Patrick Moore echoed this reasoning. "Venus is closer to the Sun than we are," Moore wrote in 1961, "and auroræ there may be expected to be stronger and more consistent. …It has its drawbacks; but unless the [Ashen] Light is after all a pure contrast phenomenon, it seems to be the best explanation."

That would be indeed the case if Venus had an intrinsic magnetic field like Earth's. Presuming, then, that Venus had a magnetic field that was probably stronger than our own, de Heen eliminated the other possibilities. In the end, he appealed to the episodic nature of auroral displays on Earth as the right way to think about the decidedly inconsistent appearances of the Ashen Light throughout history:

> We see that the first four hypotheses account to a certain extent for the luminous phenomenon; but why does not it happen regularly? This is an objection that can not be answered. … It is ascertained by the testimony of several trustworthy observers that Venus has sometimes been seen entirely, even in broad daylight, in positions in which at most only half could have been perceived. All that remains is the hypothesis which attributes the phenomenon to the presence of polar auroras, and this one is most probable.

As we saw in Chap. 2, there *is* a weak correlation between the phase of the solar cycle and the timing of Ashen Light observations recorded through history. It seems at first a plausible way to explain what's going on, but this theory depends crucially on one thing: it assumes that, like Earth, Venus possesses a (semi-)permanent magnetic field. As late as the 1950s, it seemed reasonable to assume that it did: the two planets are comparably massive, made of essentially the same materials, and must have incorporated the same amount of molten matter in and around their respective cores. Some even wondered whether the high surface temperature on Venus and its high atmospheric carbon content weren't clear indicators of volcanic activity on a grand scale, which would indicate a geologically living world. If anything, a hot, active

Venus ought to have a *stronger* field than Earth's, with correspondingly more intense aurorae.

The Dutch astronomer and planetary scientist Jakob Houtgast (1908–1982) wrote about the prospect of a magnetic Venus in 1955, inferring that its magnetic environment might even interact with our own. Houtgast reasoned that if Venus retained a permanent magnetic field, "it would at inferior conjunctions influence the corpuscular streams from the sun to the earth" and that "we should expect a variation, probably a decrease, of the corpuscular intensity, which could reflect itself in the international daily magnetic character figures." In his model, the presence of a "magnetotail" behind Venus could deflect solar wind particles around the Earth, lowering geomagnetic activity on the days around inferior conjunction. By examining geomagnetic data from a total of 44 inferior conjunctions between 1844 and 1953, Houtgast found "a marked decrease in the magnetic activity from 7 days before to 1 day after conjunction." From this he inferred a Venus magnetic field strength of about five times the intensity of the Earth's field while cautioning that "these considerations are, of course, much too simple to represent the real situation."

Hopes were therefore high when the American *Mariner 2* spacecraft visited the Venus vicinity, making the first-ever measurements close to the planet in December 1962. Its on-board magnetometer detected a magnetic field strength almost 10,000 times *weaker* than Earth's; in fact, this strength is much more like the background field throughout the inner Solar System than scientists expected. Ultimately, only fields stronger than about 10% that of Earth's magnetic field could be ruled out definitively. More modern spacecraft results peg the upper limit for the strength of the magnetic field at its surface around 10^{-4} Gauss or about 5,000 times weaker than Earth's field.

Once this was established, researchers asked why Venus ended up in such a different condition than Earth in this respect. They instinctually guessed that it had to do with Venus' very slow rotation period of 243 Earth days: if the interior of Venus is still molten, then it appears that the material that might conduct electric currents to generate magnetic fields just doesn't swirl around fast enough. Alternately, it may be that the interior has already cooled to the point where it has "frozen out," its temperature dropping below the melting point of iron. Whereas heat escaping the solid inner core of Earth drives convection currents in the molten outer core, the core of Venus could already be solid throughout, killing whatever dynamo Venus may once have had.

Newer ideas point to the possibility that Earth and Venus may have originally *both* lacked a permanent magnetic field and that Earth's convection was kickstarted by turbulent mixing of its interior due to the impact of a

Mars-sized object very early in our planet's history, driving off material that condensed into our Moon. Without a comparable impact to shake things up internally, the deep layers of Venus stratified in a way that frustrated the transport of heat from the core to the surface, preventing a dynamo from ever setting up.[7] Whatever the cause, the failure to detect a strong magnetic field around Venus isn't the fault in our ability to detect it; rather, it's because such a strong field just isn't there. As a result, any true (i.e., magnetically driven) aurorae on Venus must be so insignificant as to be an absolutely negligible potential source of the Ashen Light.

However, there's a postscript here that might lead in the right direction: not all magnetic fields are intrinsic, nor are they permanent. It is possible to briefly induce a field without the help of the geophysical machinery involved in generating a dipole field like the Earth's. Analysis of *Pioneer Venus Orbiter* data in the early 1990s indicated that interactions between the solar wind and the ionized upper atmosphere of Venus could cause the formation of a temporary magnetic environment around the planet whose strength was tied to the intensity of the wind.[8]

More recently, results from the European *Venus Express* mission confirmed the existence of a magnetotail behind Venus, "symptomatic of magnetic reconnection, a process that occurs in Earth's magnetotail but is not expected in the magnetotail of a nonmagnetized planet such as Venus."[9] In magnetic reconnection events, open magnetic field lines come together to form closed loops, causing an exchange of energy between a magnetic field and plasma with which it interacts. The observation of reconnection in the Earth's permanent magnetotail suggests that the same process might be at work in a temporary Venusian magnetotail.

In 2006, scientists detected plasma flowing toward the planet from behind Venus (as seen from the Sun) while they simultaneously observed the weak magnetic field measured by the magnetometer aboard *Venus Express* suddenly flip direction. Seeing those two events more or less simultaneously is a strong clue that reconnection had taken place. So as the solar wind sets up a temporary magnetic field in the vicinity of Venus, sculpting it into a long, flowing tail behind the planet, it also stores energy in that structure that can later be

[7] For recent work on this idea, see S. A. Jacobson, et al., 2017, Formation, stratification and mixing of the cores of Earth and Venus, *Earth and Planetary Science Letters*, 474, 375–386.

[8] J. G. Luhmann, 1991, Induced magnetic fields at the surface of Venus inferred from pioneer Venus orbiter near-periapsis measurements, *Journal of Geophysical Research*, 96, 18831–18840.

[9] T. L. Zhang, et al., 2012, Magnetic Reconnection in the Near Venusian Magnetotail, *Science*, 336(6081), 567–570.

released into the plasma. That, in turn, could create the same sort of physical conditions that help light up the terrestrial aurora.

That said, it wouldn't really be what we would call "aurora" on the Earth, in that the physical processes involved aren't quite the same. It also means that any light that resulted from the deposition of energetic charged particles into the upper reaches of the Venusian atmosphere wouldn't be confined to the planet's polar regions but rather would be equally likely at any latitude. This idea is attractive as a means of explaining the Ashen Light as an objectively real phenomenon. We'll return to this part of the story later in Chap. 7.

'Extended Twilight'

Looking toward other simple explanations for the Ashen Light, Ellen Mary Clerke wrote that

> Professor [Hermann Carl] Vogel, of Berlin, who himself saw part of the night-side of Venus, in its semi-obscurity in November, 1871, ascribed its visibility to a twilight effect caused by a very extensive atmosphere. The light thus transmitted to us by aerial diffusion and giving the ashen light, is reflected sunlight, while that sent by the luminous arc on its edge is direct sunlight, refracted, or bent round to us, from behind the planet.

She added, in typically flowery nineteenth-century prose, "The silver selvedge of the dawn edging the dark limb may consequently be the brightest part of the broken nimbus that then seems to surround her."

Vogel's idea was that a dense Venusian atmosphere might scatter sunlight so far beyond the line dividing day from night that, under the right observing conditions, it might be seen over much of the planet's night side. The "luminous arc" reported by observers as early as Giovanni Riccioli in 1643 is then interpreted as daylight reaching similarly from just beyond the edge of the planet's disc, its glow amplified by the limb brightening effect.

Patrick Moore batted down this idea: "the [Ashen] Light seems to vary in intensity, which would hardly be so were it due to general twilight or to illusion." Moore's criticism was based in part on his own observations of the Ashen Light as early as 1953, which motivated him to comment that "the variations are frequently abrupt," indicating shorter timescales than would be involved in a twilight effect unless the atmosphere of Venus was somehow unusually clumpy and wind speeds particularly high. While wind speeds of up to 700 km per hour are inferred at mid-level altitudes, the Venusian atmosphere is too uniform to explain rapid variations in a twilight-induced Ashen Light.

Light filtering through and around the atmosphere of Venus was revived as an idea by David S. Brown in 1970—with a twist. Brown (1927–1987) was what might be described as a very well-informed amateur astronomer whose career was spent designing telescopes for the firm of Sir Howard Grubb Parsons at Newcastle upon Tyne in northeast England, but he made and published many astronomical observations in the 1960s and 1970s. Citing evidence from the Soviet *Venus 4* and American *Mariner 5* spacecraft results indicating "strong refraction effects in the atmosphere," Brown argued that the Ashen Light might result from sunlight refracted through the atmosphere to the night side. The mechanism is the same as the one that produces mirages, commonly seen on hot days when looking at road surfaces near the local horizon: heated air rising from the road is much less dense than the cooler air above it, and the tendency of light to bend as it travels through a medium like air depends on the density of that medium. Light from the sky traveling toward the road is refracted upward, and the observer at a distance sees what looks like water lying across the road, directly reflecting the sky light.

Brown reasoned that if the atmosphere of Venus was well-mixed from bottom to top, it would then be isothermal (having the same temperature throughout). The result would be a gradient in the density of the atmosphere with altitude that Brown concluded should cause sunlight to bend with a radius of curvature comparable to the radius of the planet itself (Fig. 5.3). He computed a height for the volume in which light is refracted around Venus only three-tenths of a kilometer above the clouds. But he also suggested that a second mechanism exists: a more strongly refracting layer *below* the cloud tops. "If such a layer exists," Brown wrote, then "light diffused from the cloud in a direction nearly parallel to the planet's surface will be refracted to illuminate the cloud base in the opposite hemisphere." Furthermore, if that light were sufficiently intense after its flight through the atmosphere, it might rattle its way up out of the clouds and become visible from Earth.

It seemed promising at first. Citing spacecraft data showing an atmosphere that was *not* isothermal, Brown claimed that "temperature variation with height may give rise to an even stronger refractor effect." The atmosphere would then function like a waveguide, in which electromagnetic waves are essentially trapped between layers of different properties such as alternating layers of warm and cold gas. But on carrying out the calculations, one finds that the effect is still too weak. Using earthshine seen on the night side of

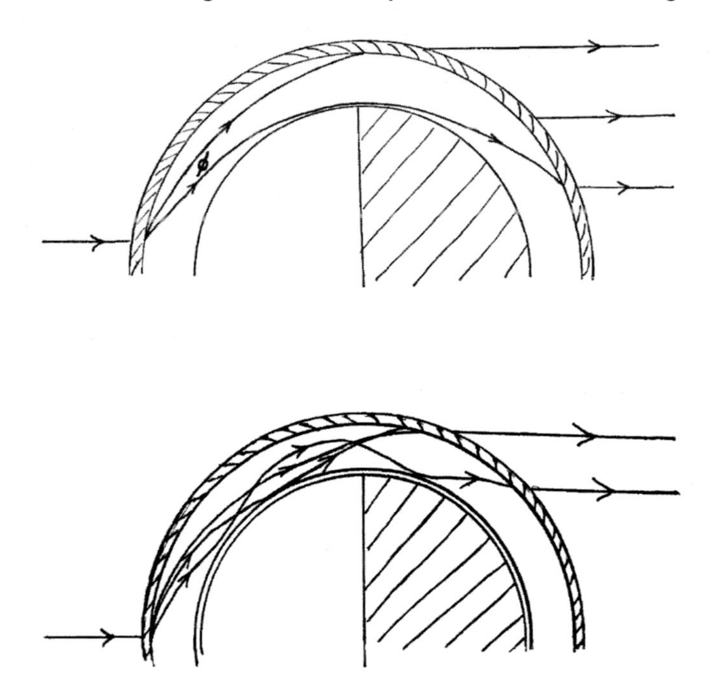

Fig. 5.3 Figures 2 and 3 from David S. Brown's 1970 *Journal of the British Astronomical Association* paper titled "The Ashen Light—A Refraction Effect?" The figures illustrate two versions of his model for the propagation of light from the day side (left) to the night side (right) of Venus through "a strongly refracting layer below cloud level." Figures reproduced with the kind permission of the British Astronomical Association and *JBAA*

the Moon as a reference, Brown calculated that the brightness ratio between the Earth-lit and Sun-lit parts of the Moon is about 1.5×10^{-4}. However, he concluded from historical accounts that the Ashen Light is certainly "more difficult to observe than earthshine," so that his ratio can only be considered an upper limit.

But Brown's model fails to account for two facts: first, knowing what we do now about the properties of the atmosphere of Venus, the attenuation of light as it refracts around the planet doesn't leave enough once it reaches the night side for it to be visible from Earth. And, second, even if the atmosphere cooperated, its opacity would have to vary both significantly and relatively quickly in order to explain the episodic nature of the Ashen Light. It therefore seems safe to discard the notion of a purely optical effect in the Venusian atmosphere.

Surface or Atmospheric Phosphorescence

Other ideas about the source of the Ashen Light appealed to some physical process, either in the atmosphere or on the surface of Venus, in which the material substance of the planet is self-luminous. "This was the idea of Sir William Herschel, [Karl Ludwig] Harding, and partly [Johann Hieronymus] Schröter," Adalbert Safarik noted in 1874.[10] Herschel seems to have first suggested this concept in reaction to the lack of a moon of Venus to produce an effect like earthshine. "The unenlightened part of the planet Venus has also been seen by different persons," Herschel wrote, "and not having a satellite, those regions that are turned from the sun cannot possibly shine by a borrowed light; so that this faint illumination must denote some phosphoric quality of the atmosphere of Venus."[11]

A decade later, Harding and Schröter both reported seeing the Ashen Light on the same night from their observatories several hundred kilometers apart in Germany. Both believed that their perception revealed a real physical phenomenon that indicated something important about the presumed atmosphere of Venus. Schröter knew about Herschel's interpretation, as well as his ideas about a "peculiar" light intrinsic to all heavenly bodies:

> Dark objects, according to the proportion of their mass and attractive force, draw upon some essential component that is changed into light. It is more probable that both are at the same time the cause of such dull light.

Regardless of the source, Schröter was "inclined to consider this dim light in the night side of Venus to be the effect of a phosphorescence, according to which this planet has the ability to develop so much light from its body that thereby its hemisphere at night, even from such a considerable distance, could still be seen." As we saw in Chap. 4, Schröter described the terrestrial atmospheric phenomenon we would now call airglow.

In his analysis, Harding considered and rejected the earthshine hypothesis, judging that such an explanation would render the Ashen Light visible on a much more regular basis than it actually was. Appealing to the "peculiar light" theory, Harding considered himself "therefore inclined to consider this twilight in the night of Venus the effect of some phosphorescence, according to which

[10] He expanded on these ideas in "Über die Sichtbarkeit der dunkeln Halbkugel des Planeten Venus," from *Sitzungsberichte der Böhmischen Gesellschaft der Wissenschaften in Prag*, 243.

[11] 1795, On the Nature and Construction of the Sun and Fixed Stars, *Philosophical Transactions of the Royal Society of London*, 85, 51.

this planet has to generate so much light from its surface that its nighttime hemisphere, even from such a distance, still can be seen."

The idea of some sort of innate luminescence of Venus works in favor of explaining the Ashen Light as long as the underlying physics producing that light depends on the condition of the carrier medium. In other words, Harding suggested something as simple as weather, with which Earthbound observers are very familiar, could exert a modulating effect that explains why we don't see the Ashen Light all the time: "If this phenomenon is the effect of a phosphorescence," he wrote, "a certain disposition of the atmosphere or of the surface of the planet is necessary in order to produce it, and the latter may fail for several years due to the coincidence of various circumstances but then in another period appear for a time."

Adalbert Safarik had some critical words in response to the notion, prevalent in the early nineteenth century, that material substances could sometimes glow for no obvious reason. Of the connotations of "phosphorescence," he wrote,

> it does not appear clearly whether they [Herschel, Harding and Schröter] understood the word in its modern sense, meaning substances which absorb sunlight and emit it in darkness without being chemically changed, or whether they included under that name, like all the elder physicists, slow combustions also, like that of phosphorus and rotten wood,[12] which in modern terminology do not belong to true phosphorescence. In both cases it is difficult to imagine the whole surface of the planet to be covered with such substances as sulphide of strontium, diamond, phosphorus, or rotten wood.

Maybe, instead, luminous surface material on Venus is ephemeral, even seasonal? This might explain why the Ashen Light was only ever seen intermittently, but in some cases it was reported by many observers over short periods of time. This explanation also conveniently removes the need for the surface of the planet to be everywhere constantly covered with some exotic substance.

Such an irregular environmental explanation was proposed by Franz von Paula Gruithuisen. As recounted previously in Chap. 4, in 1825 Gruithuisen suggested, on the basis of long spans of time between reported sightings of the Ashen Light to date in his era, the ignition of "large luxuriant forests" on the temperate surface of Venus by a race of intelligent beings to mark the occasion of the ascent of a new monarch to the throne. Such fires would burn for days

[12]Known colloquially as 'foxfire,' a faint blue-green light can be produced by certain species of fungi growing on decaying wood. This type of bioluminescence results from luciferase, a type of oxidative enzyme that emits light when it reacts chemically with luciferin, a small-molecule substrate. Not uncommon in the biological world, a luciferase variant is what powers the familiar glow of fireflies.

to weeks at a time before being extinguished, and variations in their intensities even over a span of hours might explain short-term variability of the Ashen Light. Of course, knowing as we do now that no such beings exist, much less forests for them to burn down, this fanciful explanation clearly isn't right.

Instead of a solid surface covered in some putative luminous substance, what about a light-emitting ocean? That was the contention of the German astrophysicist Johann Karl Friedrich Zöllner, held by Safarik as "a competent authority in photometric matters," who argued that the intense brilliance and apparently featureless disc of Venus were indicative of specular (mirrorlike) reflections from a very smooth surface such as an extensive ocean. Nineteenth-century scientists could scarcely imagine a sea *not* comprised of liquid water, so there was not a great leap of logic required to start at the inference of oceans and end up at a planet overflowing with life. "Should Venus be in a geologically less advanced state, viz., less cooled than our globe," Safarik wrote, "a supposition rendered not improbable by her considerable size and her nearness to the Sun, then the present condition of Venus would be analogous to that of the Earth in the Jurassic period, when large isolated islands were bathed by immense seas, blood-warm, and teeming with an abundance of animal life difficult to be conceived."

So maybe it was (bio)luminescence after all. Living things in oceans implied the likelihood of bioluminescent organisms "shown not unfrequently by our tropical seas, [giving] us some idea of the intensity which this magnificent phenomenon could acquire under such unusual circumstances." A more or less uniformly distributed, glowing ocean would also account for the observed limb brightening of the Ashen Light, and Safarik went as far to suggest that the kind of animal responsible for giving off the light could be determined by spectroscopy: "The spectroscope will be able to decide between Professor Zöllner's hypothesis and mine," he posited.

The suggestion was treated with due skepticism even in its own time. Ellen Mary Clerke wrote in 1893 that "the phosphorescence of the Aphroditean oceans, warm and teeming with life, as they are held to be by Zöllner, was advanced as an explanatory hypothesis, with scarcely more plausibility, by Professor Safarik." And, to little surprise, spectroscopy of Venus has not once revealed the slightest trace of light emission attributable to bioluminescent substances.

Thermal Emission

If it were not the surface of Venus emitting light seen through its thick atmosphere, maybe the atmosphere *itself* was so hot as to be faintly glowing? Like the filament in an incandescent light bulb, heating any medium to a sufficiently high temperature results in that material giving off visible light. And just like an incandescent filament, the higher the temperature to which it is heated, the greater fraction of the emitted light that falls within the part of the spectrum to which human eyes are sensitive. Safarik described an idea offered by Johann Wilhelm Pastorff, known mainly for his observations of the Sun, "who supposed the atmosphere of the planet to be large and self-luminous." This seems to fail as well, given that the Ashen Light should always be visible, which Safarik noted, "is positively not the case."

The temperature profile of the Venus atmosphere varies with altitude somewhat like that of the Earth: warm at the surface, dropping to a minimum tens of kilometers up, and then warming again as the absorption of sunlight increases. However, the density of gas at such heights above the surface is far too low to support emission of considerable radiation in the visible spectrum. The visible light we see from Venus is therefore attributable essentially entirely to sunlight reflected from its cloud tops and not atmospheric emission.

As it turns out, there *is* electromagnetic radiation escaping from the atmosphere of Venus, but it originates on the planet's surface. Being nearer to the Sun than the Earth we would expect its surface to be warmer, but a runaway greenhouse effect due to the composition of its dense atmosphere amplifies the difference. The average surface temperature of Venus is around 460 °C, or twice as hot as a typical kitchen oven, which means that it should shine most brightly in near-infrared light whose wavelength is around 4 millionths of a meter (four micrometers or "microns"). And indeed, images made in near-infrared light (Fig. 5.4) show faint emission from the planet's night side, which has been identified as surface thermal emission transmitted through the atmosphere. The greenhouse is pretty good, but it appears to be leaky; still, Venus could be losing substantial near-infrared light without also losing a lot of heat, as calculations show that the process lowers the equilibrium temperature of the surface by less than 1° over very long periods of time.

Venus expert Frederic Taylor, Emeritus Halley Professor of Physics at Oxford University, favors this scenario as the Ashen Light mechanism:

> The temperature at which the eye sees an object start to glow dull red is generally quoted as being about 1,000 degrees on the Kelvin scale. Venus's surface is more than 200 degrees cooler than this, but for someone with good eyesight in a

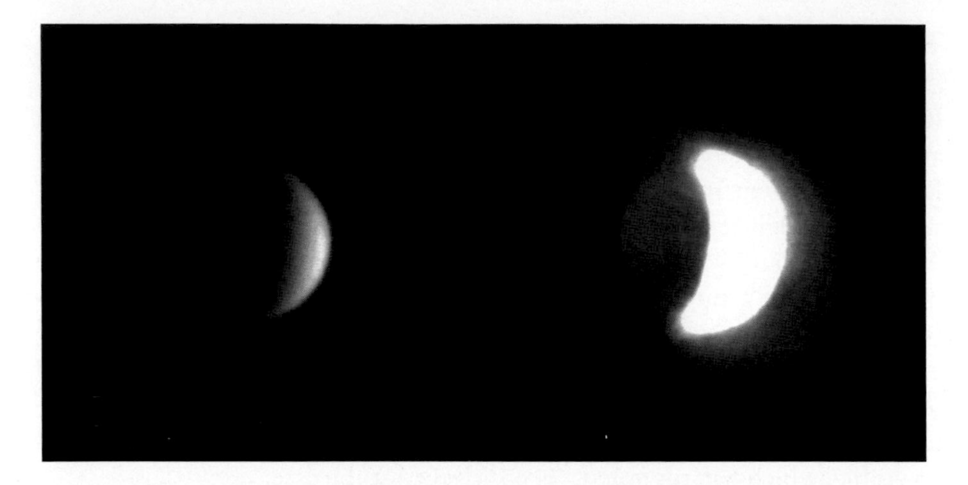

Fig. 5.4 The Italian astrophotographer Giuseppe Donatiello captured these views of Venus from Oria, Italy, on the night of June 30, 2015. Both images show the planet as seen through a germanium filter, which transmits near-infrared light with a wavelength of one micron (1 μm). The image at left is a short exposure, suggesting the visual appearance of Venus at wavelengths to which the human eye is sensitive. A longer exposure (right) overemphasizes the bright crescent but brings out the faint light due to thermal emission on the planet's night side. Images courtesy of the photographer

> completely dark room, this may be enough. If so, and we imagine ourselves standing on the surface of Venus at midnight and looking around, we would see the rocks around us glowing dimly in the gloom.

However, Taylor admits the shortcomings of the idea, "as there do not seem to have been any proper experiments done" to establish the lower limit of human visual detection of low-temperature thermal sources. "It may depend on the individual," Taylor continues, "as some eyes are more sensitive to the longer visible wavelengths than others and can see a short way into what for most of us would be the infrared. This could explain why some otherwise good observational astronomers say they can never see the Ashen Light."

On the other hand, the long wavelengths of light escaping the atmosphere of Venus might well explain why many astronomers who believed they sensed the Ashen Light through their telescopes reported its color as dull red, "coppery," or "rusty." And if the intensity of this light were near the threshold of telescopic visual detection around the time Venus was closest in its orbit to us, and naturally less intense when it was further away, it may well explain why the Ashen Light has been much more commonly reported historically near inferior conjunction rather than far from it.

Is this radiation sensed through telescopes as the Ashen Light? Probably not, for two reasons: first, like every other explanation considered up to this point, it doesn't fully account for the episodic nature of the Ashen Light; it should, in principle, be visible all the time. However, like the phosphorescence theory, the amount of surface thermal radiation reaching Earth could well be modulated by changing conditions in the Venus atmosphere that made it more or less opaque to infrared light. We know that sulfuric acid clouds in the lower atmospheric layers of Venus are efficient at scattering—that is to say, redirecting the paths of—near-infrared photons, but not absorbing them. Their random and changing distribution might explain why some Ashen Light reports indicate a patchy or mottled appearance.

The other reason why the infrared thermal emission explanation probably isn't the right one is that the human eye is terrible at sensing light of wavelengths longer than about 0.7 micron and has essentially zero response to light by the time wavelengths approach one micron. The situation is made worse at low light levels, since the sensitivity of the eye shifts to even shorter wavelengths as an efficiency switchover between different types of light-sensing cells in the retina takes place. Surface thermal emission just doesn't seem to generate enough light that the telescopically aided eye can detect. "When Venus is very near [inferior] conjunction, when at just a few degrees of separation from the Sun, the intensity of the heat emission at 0.7 micrometers is some 30,000 times weaker than the brightness of the sunlit sky," Bill Sheehan and coauthors wrote in 2014, "far too weak" to be seen above the sky level. The situation might be improved by observing Venus when it's further away from the Sun in the sky and therefore can be seen against a darker background. In that case, however, the brightness of the source decreases faster than the brightness of the background, making positive detection that much less likely.

One way to make this work is if the human eye is more sensitive to near-infrared light than we think. When navigational radar was invented during the 1930s and put into practical use during World War II, some researchers began to wonder whether airplane pilots could sense long-wavelength light if it were presented to their eyes at sufficient intensity. Research done in that era showed that for wavelengths of light longer than about 0.8 micron, where the eye otherwise should show no response to light, certain faint-light receptors in the retina become more sensitive than other cells that give color discrimination under bright lighting conditions, and these conditions lead to positive detections among human test subjects.[13] Further studies in the 1960s and 1970s revealed that the eye could, under certain circumstances, sense light with a wavelength as long as 1.355 microns.[14]

In December 2014, Grazyna Palczewska and coworkers published a paper in the *Proceedings of the National Academy of Sciences* entitled "Human Infrared Vision Is Triggered by Two-Photon Chromophore Isomerization,"[15] which, they claimed, "resolves a long-standing question about the ability of humans to perceive near-infrared radiation and identifies a mechanism driving human IR vision." The biochemical mechanism that Palczewska and company proposed to explain why the eye can sometimes sense such long wavelengths relies on a quantum mechanical trick in which two near-infrared photons happen to land on the same receptor in the human retina simultaneously. If those two photons each have a wavelength of one micron, they will deliver the same amount of energy as that involved in the absorption of a single photon with a wavelength of half a micron, because the energy carried by a photon is inversely proportional to its wavelength. Receptors in the human eye are sensitive to light with that wavelength, and absorption of that much energy leaves a receptor molecule called rhodopsin in an energetically excited state, triggering a nerve impulse that the brain interprets as "green." That, the authors explained, was why their test subjects "perceived low-energy IR laser radiation of 1060 nm [1.060 microns] as a pale greenish light"—again, not inconsistent with some reports of the color of the Ashen Light.

However, the radiant power of the laser light source used in their laboratory experiments on human volunteers, about one milliwatt, is far greater than what we would expect given eyewitness descriptions of the faintness of the Ashen Light. Additionally, we don't know what the lower limit is for the intensity of infrared light that can still trigger the "two-photon" mechanism. This is all further complicated by the fact that the researchers exposed their test subjects to pulsed laser light, meaning that the light was delivered in very short doses between which no light was received. In fact, the shorter the pulse, the more likely it was a subject reported sensing the infrared light; it seems that the process needs a "relaxation" time when the receptor *isn't* being irradiated in order to work. But Ashen Light observers report that the phenomenon is steady

[13] D. R. Griffin, R. Hubbard, and G. Wald. 1947. The sensitivity of the human eye to infra-red radiation. *Journal of the Optical Society of America* 37(7), 546–554.

[14] V. G. Dmitriev, et al. 1979. Nonlinear perception of infra-red radiation in the 800–1355 nm range with human eye. *Soviet Journal of Quantum Electronics*, 9(4), 475–479.

[15] *PNAS*, 111, 50, E5445–E5454.

in intensity over minutes to hours, undercutting the notion that the eye really is particularly sensitive to the surface thermal emission.

On balance, the emission of infrared light from the night side of Venus can't be ruled either in *or* out as a means of explaining the Ashen Light. At best we might say that certain circumstances may align that make it more or less likely and that this itself is worthy of additional investigation.

Surface Ice

Robert Barker, a British amateur astronomer and prolific lunar observer of the early twentieth century, made a curious suggestion before the November 1953 Ordinary General Meeting of the British Astronomical Association.[16] Barker argued that the reflectivity of the sunlit portion of Venus' disc—its "albedo"—was so large that "it seemed incredible" that "a cloud-covered planet could attain so high a value." Maybe something else was going on. And, rather than telling us something about the atmosphere of Venus, Barker argued, the Ashen Light may be telling us something important about the planet's *surface*.

At the time, it was believed that the rotation of Venus was "tidally locked" to its orbit such that the same face always pointed at the Sun. This belief stemmed from the fact that its rotation period was then unknown; only a few illusory markings, whose regular appearances could be used to make rotation timings, were ever seen on its surface by telescopic observers. A tidally locked (but cloud-covered) Venus would rotate so slowly such that Earthbound observers wouldn't perceive any rotation at all. Assuming that the Venus atmosphere was relatively thin, the great difference in temperature between the day and night sides should set up a powerful global wind circulation.

Barker saw a problem, if indeed that was what was going on. "I know very little of physics," Barker told the assembly of astronomers at Burlington House in Piccadilly, London:

> and that is why I ask physicists here present to explain how a planet eternally facing the Sun can retain a cloud-laden atmosphere. One face must be baked for countless ages, and the other, chilled by everlasting night, must produce indraughts of air from the cold to the hot side of great power. A tornado of air of varying temperatures must be eternally encircling Venus, and this aerial circulation would remove all moisture from the sunward face to deposit it as

[16]Recorded in the *Journal of the British Astronomical Association*, 64, 60–62, ed. E. A. Beet and D. A. Campbell.

> ice on the night side. This can explain the ashen light (or *lumière cendrée*) of the rearward face when presented to us at inferior conjunction, a feature I have observed.

Barker ended his presentation asking for "mercy and not sacrifice" from members of his audience "who have keen objections to any suggested alternation to the planetary data which, to them, are axiomatic."

Fortunately for him, as memorialized in the proceedings of the event, the other BAA members at the meeting handled Barker's ideas gently but skeptically. William Herbert Steavenson, a physician and surgeon by training, spoke up, noting that he thought the idea was rather similar to one promoted years earlier by the late American astronomer Percival Lowell, who believed he saw dark, radially oriented streaks on Venus. These were, Steavenson said he remembered reading, "the effects of air currents, drawn with great velocity from the dark side toward the subsolar point."[17]

Arthur C. Clarke, the science fiction author and futurist, offered that the high albedo of Venus might involve fluorescence by some substance in the atmosphere or on the surface, although others (the planetary photographer Horace E. S. Dall chief among them) discounted the idea, noting that exotic materials were not required to achieve high albedo. A.P. Lenham, a physicist at the Royal Military College of Science, objected to Barker's idea on the grounds that it still didn't explain the source of the reflected light. The problem remained as it did with any theory involving direct reflection of light, which was to find a source bright enough to yield a detectable signal when the reflected light arrived on Earth.

Patrick Moore was absent from that particular BAA meeting, but he criticized Barker's theory in 1961 on the grounds that it "involves acceptance of the 225-day rotation period and also of Lowell's contention that we are looking at the actual surface of Venus instead of the upper part of an atmosphere—which seems unlikely." Moore was less reserved two decades later: "Conditions [on Venus] do not seem to be very well suited to the formation of ice sheets!"

In the era before spacecraft sent to Venus from Earth delivered close-up impressions of our neighboring world, it was not absurd to suggest that conditions there were similar to those here at home. The reason for this, Richard Baum wrote in 2007, was the Victorian view of Venus:

[17] Sheehan (1988) makes a strong case that the apparent "spokes" Lowell claimed to have seen on Venus may have been nothing more than defects in Lowell's eyes, obliquely illuminated and seen in shadow against the bright disc.

> In the drama of human opinion, Venus had undergone a scene change. … What has once seemed idyllic, a place of dense jungles and luxuriant life, had been transformed into a barren, windswept desolation of harsh extremes: 'One face baked for countless aeons, and still baking, backed by one chilled by everlasting night, while both are still surrounded by air.'[18] Venus, it appeared, was a world disinherited from all the gifts of Nature — a frightful habilitation, deserted and silent save for the howling of the wind though the high mountain passes.

Nineteenth-century astronomers saw all manner of light and shade on the disc of Venus, including bright "hoods" near the cusps of the Venus crescent, corresponding presumably to its polar regions. Finding analogs in Earth and Mars, some observers interpreted them as ice caps seen either through or protruding above the clouds. Given that the Red Planet hosted polar ice caps that seemed to wax and wane with the Martian seasons, it was not unreasonable to presume that Venus was possessed of the same.

The polar hoods were first recounted by Franz von Paula Gruithuisen, ever the "source of a good story, exemplifying the eccentric fringe of astronomy," who we met in Chap. 4 when he proposed an eyebrow-raising theory about intelligent life on Venus. Gruithuisen first saw bright spots in the polar regions of Venus in December 1813, noting what Baum described as "a curious swollen look to the southern horn of the crescent." Given their evident permanence, according to his observations, Gruithuisen decided that they must be great rafts of ice surrounding the poles. The notion took root and by late century was fairly widely accepted. In 1889, George Frederick Chambers wrote that there was "no prima facie reason why it should not be well founded: indeed rather the reverse."[19] Furthermore, since the spots seemed immobile, their persistence was taken as evidence for the tidal locking of Venus, creating a sense of circular logic as to how the ice came to form in the first place.

Belief in ice caps on Venus was seemingly rewarded in the 1920s when the first photographs of Venus obtained in ultraviolet light validated the objective reality of bright spots in the polar regions of the planet. Further images obtained by spacecraft flying by and orbiting Venus in the 1960s and 1970s confirmed that telescopic observers on Earth before the turn of the twentieth century *did* see real features in the clouds of Venus. But they were just clouds, not ice. Whatever hopes remained for ice on Venus melted in

---

[18] Quoting Percival Lowell, 1910, in *The Evolution of Worlds*, New York: MacMillan, 80.

[19] *A Handbook of Descriptive and Practical Astronomy*, Oxford University Press (4th ed.), 1, 101. Chambers (1841–1915) was an English barrister and author who switched to law after initially studying engineering. The 600-page first edition of his *Handbook* was published when Chambers was only 19 years old, undergoing an expansion to three volumes in subsequent editions.

the 1950s when measurements of the microwave spectrum of Venus implied a surface temperature of hundreds of degrees.

Ironically, the notion of "ice" on Venus was vindicated almost 60 years after Robert Barker advanced his theory to the British Astronomical Association. In October 2012, European scientists studying data from the *Venus Express* spacecraft found evidence for a surprisingly cold layer of the Venusian atmosphere. It is so cold, in fact, that it's possible under the conditions in this layer for carbon monoxide (CO) molecules to "freeze out" as ice crystals. Some 125 km above the surface, the temperature of the atmosphere plunges briefly to −175 °C. In stark contrast to the hellish conditions deeper in the atmosphere, it's possible that a dry ice "snow" just briefly falls there. Given the temperature at the surface, the CO is liberated as a steamy vapor long before it gets anywhere near the ground.

But what about "snow" of another kind? The US *Magellan* spacecraft observed something odd about the reflectivity of certain parts of the planet in radio energy. During its 4-year mission in the early 1990s, the spacecraft mapped Venus through its thick clouds in exquisite detail by bouncing radio waves off its surface. Certain parts of the planet appeared to be much more radio-reflective than others, and the shiniest spots of all turned out to be among those highest in mean elevation. As *Magellan*'s radar beam swept over high points such as Maxwell Montes and Aphrodite Terra, it illuminated what is now thought to be "Venus snow," a millimeters-thin layer of sulfides of lead and bismuth that has precipitated onto these surfaces directly from the atmosphere.

The favored mechanism to explain the metallic snow is one in which metallic compounds from the surface form a vapor at the high pressures and temperatures of the lowlands and are transported through the thick lower atmosphere where they condense out as solids in higher (and therefore cooler) places. Alas, this explanation for the Ashen Light, too, falls short. While the mineral deposits in the highlands of Venus may be highly reflective to radio waves, they are far too dull at the wavelengths of optical light, and their locations are too few in number, to explain any visible light seen toward the planet's night side.

Scattered Light and Diffraction Effects

Lastly, some of the early thinking about the Ashen Light considered the possibility that perhaps the very simplest reason why an observer looking at Venus through the eyepiece of a telescope would sense light where it shouldn't otherwise be on the planet's night side is because the telescope itself put the light there. To explain why that would be the case requires a slight diversion into the details of how telescopes bring us detailed views of faraway things.

Whether using mirrors, lenses, or a combination of the two, all telescope designs have in common one particular characteristic: they direct the light collected over a large surface area to a focus over a much smaller surface area. In the case of an eyeball looking into an eyepiece, that surface area is the size of the pupil in cross section, so the telescope in effect vastly increases the light collecting power of the eye. Astronomical sources of light are so far away that their rays arrive essentially parallel to one another, modified by distortions that the Earth's atmosphere imposes on them during their brief flight through the air in the last split second of their travel time.

But let's suppose, for a moment, that the Earth has no atmosphere and the rays entering a telescope are truly parallel. If all of the optical elements of the telescope, be they mirrors or lenses, are in proper alignment, then the images of stars should come to focus as tight, infinitesimally small points. The features of the Sun, Moon, and planets seen through a telescope on the airless Earth would be perfectly sharp all the way to the edge of the field of view, assuming ideal optics with no aberrations. And contrasts should be absolute: places on the sky where there were no sources of light should completely dark, an inky blackness that few among us can imagine.

Even ignoring natural sources of faint light in our own atmosphere, we still don't see such views through telescopes. In part that is because some of the light scatters internally within the telescope. Sometimes that's the result of imprecisely aligned optics, with slight tilts and displacements between the elements disrupting the parallelism of the light rays in the "train" between the first optical element and the last. Once individual light rays step off this train, they can bounce off any obstacle, changing their directions with each internal scattering until they are either redirected out of the telescope or absorbed somewhere within the system. Astronomers go to great lengths to minimize these reflections, adding carefully designed and positioned absorbers and coating lenses with materials that prevent light from being reflected from their surfaces.

As we learned in Chap. 2, the situation is further complicated by the diffraction of light from various surfaces in the optical train including the edge of the telescope's circular aperture itself. Point sources like stars are seen surrounded by faint rings of light like the ripples on the surface of a pond expanding outward from the place where a stone is dropped in, but frozen in time. For angularly extended sources of light like the planets, it's as if the visible object is composed of an infinity of point sources, each with their set of "diffraction rings." The result is a blurring of the object that would exist even if the optical train were perfectly aligned and there were no internal scatterings to speak of. The effects of diffraction and stray light in the optical paths of telescopes can add up to send light in directions that might otherwise lead visual observers to draw incorrect conclusions about what they think they see.

An explanation intrinsic to telescopes themselves also doesn't seem to account for the persistence of Ashen Light as seen by observers who make efforts to block from view the direct light emitted by the bright crescent of Venus. The observability of a faint light source is strongly impeded by glare if a much brighter source is in the same field of view; this is why we often hold out a hand at night to shield direct light from intense sources in order to better see whatever those sources illuminate. Since essentially every account of the Ashen Light has it as vastly fainter than the illuminated crescent, the potential clearly exists that it could be overlooked unless there were some ways to keep only the night side of the planet in view.

One means of accomplishing this is to place a physical barrier to that light in the field of view of the telescope. Astronomers long ago figured out that the sensible place to put such a barrier is rather close to the eye. An object placed just in front of the ocular lens, with respect to incoming light from the telescope, will be seen by the eye in focus. Placing a strip of material across the lens at this point makes an "occulting bar," intended for just the purpose of hiding a bright object—say, the brighter component of a binary star system—so that the eye can sense the much fainter light of other objects in the field. Since the bar is in focus, its shape can be adjusted according to the nature of whatever is being observed. Some astronomers doing visual work on Venus have created elaborate masks designed to hide the crescent while leaving visible as much of the night side as possible. This should significantly improve the visibility of any faint light there, and that's exactly what some dedicated observers have reported. While this approach doesn't remove the effects of diffraction, which is always present due to the nature of light itself, it does dampen the effects of scattered light in a meaningful way.

Another means of hiding the bright crescent is to put it temporarily behind something else in the foreground. The Moon makes a good screen, as on occasion it is seen to pass directly in front of Venus as seen from Earth. If this happens during a western (morning) elongation of Venus, the night side emerges from behind the Moon before the crescent; for a few seconds, only light from the night side is seen from Earth. The right combination of circumstances is rare, only arising every couple of decades, and history doesn't present any record of the Ashen Light being spotted during a lunar occultation.

David and Joan Dunham of the International Occultation Timing Association described searching for the Ashen Light from Alice Springs, Australia, during a favorable lunar occultation of Venus on October 8, 2015. The Dunhams looked carefully for indications of the Ashen Light in visual observations and image-intensified video footage taken through small telescopes but saw nothing. In a presentation at the 231st meeting of the American Astronomical Society in January 2018, they reached the conclusion that:

> the ashen light, as it was classically defined, is not observable visually or with small telescopes in the visual regime. ... To date, the authors have been unable to locate any reports of others attempting to observe this unique event. That is a pity since, not only was it interesting for an attempt to verify past observations of the ashen light, it was also a visually stunning event.[20]

In recent years, observers have taken to using digital imaging systems to try to capture any light from the night side of Venus during a morning occultation, as high sensitivity and rapid readouts of image data from modern sensors yield images with good signal and sufficient time resolution. But, again, nothing: if the Ashen Light is a real phenomenon on Venus, it hasn't made its presence known during lunar occultations. On the other hand, if it is objectively real, given its episodic nature, it's possible that it simply wasn't active on nights of occultations in years past.

Besides concerns about chromatic aberration in telescope lenses, which appeared in the first recorded eyewitness account of the Ashen Light in 1643, other issues with telescopes that might result in spurious detections were identified as problematic as early as the eighteenth century. Editing the remarks of Friedrich von Hahn for the *Berliner astronomisches Jahrbuch für 1796*, Johann Elert Bode reasoned that lens aberrations explained von Hahn's report of seeing faint light on the night side of Venus, suspecting that "the circle of confusion of the vivid light rays" radiating from the bright crescent existed only in von Hahn's telescope. In this confusion, Bode concluded, was born an illusion of light where none truly existed.

In 1914 the French astronomer André-Louis Danjon (1890–1967) proposed that haloes of violet light around very bright objects resulting from the incomplete correction of chromatic aberration in refracting telescope lenses explained most, if not all, reports of the daytime, so-called "negative Ashen Light" made by astronomers using refractors. In these cases, observers sense

[20]The Reappearance of Venus Observed 8 October 2015, American Astronomical Society Meeting #231, abstract 144.11.

that the night side of the planet is somehow *darker* than the surrounding daytime sky. Something like this might be possible if Venus were seen against a bright background around the time of inferior conjunction, such as the zodiacal light or the Sun's extended outer atmosphere. The *foreground* light of the daytime sky poses a problem in the case of the latter, and searches for any sort of shadowing effect against the corona during total solar eclipses have turned up nothing.

Danjon argued that the appearance of a dark Venus disc against the daytime sky is merely a perceptual flaw aided by a "lack of achromatism" in telescope lenses. "We do not doubt for a moment that it is purely illusory," Danjon explained, "for a slight mist [in the air], rendering the sky less transparent, did not attenuate the phenomenon."[21] The reason, Danjon argued, was the color of the aberrant light:

> The purple halo with which Venus surrounds is the sum of the violet circles relative to each of the points of the crescent.[22] It is easy to see that the intensity of the resulting halo is greater in the concavity of the crescent than on the side of the convexity. In other words, the sky, during the day, is tinged with purple inside the crescent and it looks darker. Imagination is enough to complete the illusion.

As evidence supporting his hypothesis, Danjon pointed out that some observers noticed that the use of "a yellow filter, extinguishing the violet halo, completely destroys the appearance of the dark disc."

Twenty years later, Danjon again wrote about the Ashen Light, arguing that optical effects must also explain "positive" sightings against a dark night sky. Noting that many historical descriptions have the Ashen Light brighter on the side of the disc facing toward, rather than away from, the bright crescent, he decided that achromatism of refractor lenses must also account for the nighttime appearances:

> Each point of the crescent behaves like a star: it gives rise to a circular diffusion spot. The composition of the elementary spots gives a halo much more intense in the concavity of the crescent than on its convexity. Venus having, in inferior conjunction, an apparent diameter close to one arcminute, the inner halo extends

---

[21] 1914, La Planète Vénus en 1913, *L'Astronomie*, 28, 221–232.

[22] In order to explain the observed pattern of light that accounts for the phenomenon, Danjon means that each discrete point of light along the crescent becomes the source of its own violet halo.

> at least to the center of the dark disk; it covers almost entirely this disk, when the extent of the crescent exceeds a half circumference.[23]

In the end, Danjon concluded that the Ashen Light must have "a purely optical explanation" and that the whole business should simply "disappear from the list of planetary phenomena." However, Danjon inaccurately believed that "the Ashen Light of Venus is completely invisible in reflecting telescopes." The many recorded observations with reflectors, free of the chromatic aberration of their refracting kin, can't be accounted for according strictly to Danjon's logic.

It's impossible to completely rule out this explanation without methodically examining each telescope used to make a purported Ashen Light observation, at least insofar as internally scattered light is concerned. At the same time, it seems that any optical misalignments would otherwise be more or less permanent, and thus bogus Ashen Light detections should recur every time a particular telescope was used to view Venus. The alignment of optical surfaces might change if, say, the user made adjustments to the optics to improve the situation or the optics shifted over time due to expansion and contraction of the telescope tube and other components due to cycles of warm and cold temperatures, but sightings of the Ashen Light don't follow patterns that would tend to support this idea.

Observers who say they saw the Ashen Light repeatedly don't report that the phenomenon suddenly disappears for good after their telescopes are given a good tune-up. And it's further problematic that observers using very fine refractors, without the design elements in the optical train of many reflectors that lead to significant problems with diffraction, have seen it. On balance, it appears that stray light or diffraction effects can't easily explain away the Ashen Light.

If the Ashen Light is a real, physical phenomenon and not a ghost in the machinery of telescope optics and human eyes and brains, then it must result from a particular combination of physics and/or chemistry on or near Venus itself. Weighing the various ideas proposed up to the early decades of the twentieth century, those that withstand scrutiny the best point to something having to do with the atmosphere of Venus. We will come back to both ideas in the chapters that follow.

---

[23] 1934, Sur la Pretendue Lumière Cendrée de Venus, *L'Astronomie*, 48, 370–372. Friedrich von Hahn reported that the Ashen Light effect was reduced in angular extent and considerably fainter as the Venus crescent became very thin around inferior conjunction, consistent with Danjon's explanation. See von Hahn's 1793 drawings illustrating this effect in Fig. 3.5.

6

Venus as a Knowable World: Chasing the Ashen Light into the Space Age 1900–1980

At the start of the twentieth century, astronomers had reached something of an impasse in terms of understanding the Ashen Light. Although the available documentary evidence suggests that it was then widely viewed as a real phenomenon, there was nothing close to a consensus as to what caused it. The status quo was ironically frustrated by the advance of technology: bigger and better telescopes offered no more information to visual observers about what was going on, and the photographic process failed to yield incontrovertible proof of the Ashen Light's existence.[1] The sense of frustration persisted for decades, and a certain malaise about understanding Venus failed to resolve when the first spacecraft to visit the planet sent back mostly featureless images of its bland and impenetrable upper cloud deck. Even the smallest hoped-for windows to surface views through the clouds remained stubbornly shut, and humanity's knowledge of the deeper atmosphere was limited to information encoded into the infrared light and radio waves that leaked through the otherwise opaque clouds.

Seeing little forward progress in Venus studies, some writers simply threw up their hands in helplessness. A standard entry-level text in wide circulation at that time was *Astronomy For Everybody*, published in 1902 by the self-taught Canadian astronomer and polymath Simon Newcomb (1835–1909).

[1] In fairness, photographic emulsions of the time were largely unsuited to the task of capturing the emission of faint light except as deployed on the largest telescopes then available. If the Ashen Light were as faint as many eyewitness reports suggest, it was hopeless to expect that the "slow" emulsions then available could have recorded it. For an early account of the history of astronomical photography, see Daniel Norman, 1938, The Development of Astronomical Photography, *Osiris*, 5, 560–594.

J. C. Barentine, *Mystery of the Ashen Light of Venus*, Astronomers' Universe,
https://doi.org/10.1007/978-3-030-72715-4_6

In it Newcomb mentioned the Ashen Light and reviewed some of the ideas to explain it advanced up to that point in time. Newcomb evidently didn't believe that the Ashen Light was a patently false perception among those who reported seeing it. "But," he concluded,

> it is more likely due to an optical illusion. It has generally been seen in the daytime, when the sky is brightly illuminated, and when any faint light like that of phosphorescence would be completely invisible. To whatever we might attribute the light, it ought to be seen far better after the end of twilight in the evening than during the daytime. The fact that it is not seen then seems to be conclusive against its reality.[2] The appearance illustrates a well-known psychological law, that the imagination is apt to put in what it is accustomed to see, even when the object is not there. We are so accustomed to the appearance on the moon that when we look at Venus the similarity of the general phenomena leads us to make this supposed familiar addition to it.

In the same year that Newcomb shuffled off this mortal coil, the Irish amateur astronomer and author John Ellard Gore (1845–1910) also considered the facts of the case to date and the various ideas put forth about what was going on. He ultimately decided that:

> none of these explanations are entirely satisfactory, and the phenomenon, if real, remains a sort of astronomical enigma. The fact that the 'light' is visible on some occasions and not on others would render some of the explanations improbable or even inadmissible.[3]

Around the same time, lunar and planetary astronomy was passing out of vogue in both Europe and the United States. This was in part driven by advances in engineering that enabled construction of the first large reflecting telescopes with precisely figured mirrors. Until the late nineteenth century, the world's largest telescopes were almost all exclusively those that used lenses rather than mirrors to collect and focus light. Although designs for such refracting telescopes yielded better correction for the problem of chromatic aberration discussed in Chap. 2, large, well-corrected refractors were still relatively unavailable at the turn of the century due to their high cost. Given

---

[2] It's unclear why Newcomb thought that the Ashen Light was never seen against a dark night sky, for any cursory examination of the literature in his time would have turned up many such accounts. But Newcomb seemed inclined to disbelieve the physical reality of the Ashen Light, introducing it as an "illusion to which the sight of even good observers may be subject."

[3] 1909, *Astronomical Curiosities Facts and Fallacies*, London: Chatto and Windus, 27.

the number of historical refractors still in frequent use at the time, their tendency to yield chromatically distorted images made them a target for criticism in terms of interpreting certain accounts of Ashen Light sightings.

We encountered the ideas of the French astronomer André-Louis Danjon in the previous chapter. Danjon thought that *all* reports of the Ashen Light during the daytime—it's "negative" form, in which the disc of Venus appears darker than the daytime sky—resulted from the colored haloes around bright objects imparted by chromatic aberration inherent in poorly corrected refractor objective lenses. To prove his point, Danjon noted that the use of a yellow filter at the telescope eyepiece seemed to make any appearance of a dark Venus disk seen during the daytime vanish completely. The use of yellow filters was an old astronomer's trick to compensate for chromatic aberration when the color of the object under observation wasn't important, and Danjon made clear he tested out this notion himself in his observations from Paris during the favorable evening apparition of Venus in the spring of 1913. Furthermore, the transparency of the sky didn't seem to matter, as Danjon noticed the same effect even under conditions of a "slight mist" in the air.

Twenty years later, Danjon revived his argument in reaction to an observation of the Ashen Light in June 1932 by the pioneering French astrophotographer Ferdinand Quénisset (1872–1951) at Camille Flammarion's observatory in Juvisy-sur-Orge. On the day of inferior conjunction, and with the horns of the Venus crescent extending around three-quarters of the planet's circumference, Quénisset reported that "part of the planet's disc inside the crescent was frankly darker than the surrounding sky."[4] But he also saw additional light "between the horns" of the crescent in a description that suggests forward scattering of light in the atmosphere rather than the true Ashen Light; in any case, he thought it was some sort of contrast effect unrelated to the appearance of the dark disc.

Danjon shot back that there was nothing unusual about Quénisset's apparently singular observation: "It's always how I saw the phenomenon." Whatever Quénisset *thought* he saw, Danjon asserted, it was

> irreconcilable with the usual astronomical explanation: the dark hemisphere of Venus, it is said, would stand out in silhouette against the luminous background constituted by the zodiacal light. On the contrary, it argues in favor of an optical explanation.

[4]1932, La 'Lumière Cendrée' de Venus en 1932. *L'Astronomie*, 46, 370. This appears to be the first unambiguous reference in the scientific literature to the "negative" Ashen Light.

It was, he further argued, the source of multiple logical contradictions that undermined any explanation based on an underlying physical mechanism. First,

> if the dark hemisphere stands out against a luminous mass behind it, it should be more easily visible at dusk or at night than during the day. But it is not so: the light of day seems indispensable to the manifestation of the phenomenon. When the ashen light of Venus is visible in the middle of the day, it disappears before the end of the day. The observation cannot be made at the moment of inferior conjunction, but, as M. Quénisset reminds us, the ashen light is seen at elongations greater than that of the maximum brightness (39°), when the planet can be followed nearly two hours after sunset: the dark hemisphere never appears under these conditions.[5]

Second,

> if the hemisphere emitted a true ashen light similar to that of the Moon, it should be extremely intense to manifest in broad daylight, and it would appear brilliantly in the twilight, which does not occur. The observers who thought they could see the dark hemisphere illuminated by a faint glow were the playthings of an illusion. On the contrary, the apparent darkening has an objective cause, which legitimizes the discussion that is being made here.

Danjon continued. "Third contradiction:"

> the dark hemisphere does not have the aspect that one should observe, if it were visible against a luminous background. It is not in the vicinity of the crescent that it would appear with the maximum of contrast, but on the opposite, near the dark edge of the planet, that would be itself the most apparent detail. The darkening appears degraded from this edge to the crescent. Now, the dark edge is not visible, [and] the observation of M. Quénisset leaves no doubt on this point; as for the gradient, it takes place from the crescent towards the center of the disc, contrary to expectations.

After rejecting any physical process inherent to Venus accounting for the facts, Danjon returned to his 1914 assertion that the Ashen Light was just a fault of the lens with no independent basis in reality. But he then made a

[5]As in the case of Simon Newcomb's pronouncement of the "fact that it [the Ashen Light] is not seen … after the end of twilight in the evening," it's unclear why Danjon claimed something that was controvertible by literature references available in his time, unless he either deliberately ignored observations incongruent with his hypothesis or he was simply unaware of those accounts.

crucial error of fact by claiming that "the ashen light of Venus is completely invisible in reflecting telescopes." Danjon probably didn't know that observers equipped with reflectors reported the Ashen Light as early as the eighteenth century and would continue to do so well past his own time. His idea also fails to account for "positive" Ashen Light observations when the bright crescent of Venus is masked out or otherwise hidden from direct view, in which case no violet halo should be produced.

Once the problem of chromatic aberration caused by their lenses was largely solved, refractors became very attractive for astronomers: they did not rely on early telescope mirrors whose reflective surfaces tended to tarnish easily, nor did they suffer the obstruction of light from mirrors suspended in the optical train. On the other hand, refractors often had long and unwieldy tubes requiring voluminous enclosures, and the manufacturing process that yielded workable glass reached its practicable limit with the 1.25-meter-diameter objective lens cast for the Great Paris Exhibition Telescope of 1900. Not only were such instruments virtually impossible to mount in steerable tubes, but the need to make the lenses relatively lightweight led to the engineering of a fatal flaw: their strength was not enough, relative to their thickness, to avoid drooping under the force of gravity. The achievements of refracting telescope design were the ultimate downfall of these instruments, especially as astronomers gradually demanded more and more light. For a variety of reasons, the push for bigger glass would favor surfaces that reflected, rather than bent, astronomical light.

Lunar and planetary observations benefit from large telescope apertures, but only to a point. More light enables high magnification of images in order to best see those fine features, but with it comes a steady diminution of contrast. Subtle shadings on the surfaces or in the atmospheres of Solar System objects often indicate real features that have physical explanations, so the ability to visually detect or photographically record those features was crucial to advancing planetary science in the era before direct robotic exploration was possible. Both visual observers and astrophotographers experimented with approaches such as the use of aperture stops and colored filters to overcome this effect, devising a set of strategies that are still important tools in the amateur astronomer's kit.

As fairly big and bright objects viewed though the distorting medium of the Earth's atmosphere, the Moon and planets are subject to the law of diminishing returns. Bigger telescopes provide better angular resolution, which in principle allows observers to see progressively smaller features on distant objects, but the efficacy of these systems is coupled to the nuances of atmospheric turbulence. This leads to the concept of a "sweet spot" in terms of telescope aperture in which the resolution and light-gathering advantages of size are offset by losses both within and beyond the telescope. Amateur astronomers who own telescopes of various sizes are well acquainted with the effect, often owning more than one instrument optimized for different targets: big "light bucket" reflectors to spot faint nebulae and galaxies, and much more modestly sized refractors for stunning views of Solar System objects.

Having had the opportunity to look at the Moon and planets through telescopes whose apertures ranged in size from two centimeters to four meters, I can say with confidence that the best views seem to come from remarkably small instruments between about 15 and 40 centimeters in diameter. On the other hand, the appearance of planets through very large telescopes is often disappointing—and that assumes that the brightness of such objects can be tamed to the point that one sees *anything* other than a featureless disc.

Stepping back momentarily, the real issue here is why astronomers crave bigger telescopes to begin with. It turns out that their scientific interests have long interacted with engineering and technology innovations. Sometimes necessity is the mother of invention, and science questions drive hardware development until the emergence of instruments that can adequately address those questions. At other times in history, the roles have been reversed; for instance, modern digital imaging capabilities that revolutionized the detection of light evolved from efforts to invent a new kind of computer memory chip.[6] Modes of use undreamt of by inventors occasionally (and forcefully) open new discovery spaces in other fields that change science forever.

The invention of the photographic process in the 1820s and 1830s played a role not unlike those memory chips that became the basis for astronomical charge-coupled devices in the 1970s. Permanently fixing images, like recording audible sounds for later playback, has an obvious aesthetic appeal. For one thing, photography enables what many hoped would be durable, objective views of the world and its inhabitants; after all, the photographic plate simply helps create a literal artifact of where light does and does not impinge on its surface. A picture may be worth a thousand words, but the message it conveys is of limited use without proper context and careful interpretation. Still, some nineteenth-century astronomers correctly anticipated the predominance of the photographic emulsion as the main recorder of light in decades to come, due as much to a perception of objectivity in the record as the fact that a permanent

[6]For an early history of digital imaging in astronomy, including an image that purports to be the first of an astronomical nature ever made with a proper digital imaging device, see J. Janesick and M. Blouke, 1987, Sky on a Chip: The Fabulous CCD, *Sky and Telescope*, 74, 238–242.

image enabled study of astronomical objects away from the eyepiece and under circumstances that yielded less to observer bias and fatigue.

More importantly, photography literally enabled that which the eye and brain could never by themselves sense. While the historical photographic process is remarkably inefficient by modern standards in the sense that the response of the best chemical emulsions to light is only a few percent, time exposures offered a way forward. Plate cameras mounted on the backs of telescopes could stare at the same piece of sky for hours at a time, the telescopes engineered to move in such a way that they countered the rotation of the Earth around its own axis. Instead of smearing the image across a plate, telescopes can keep the image neatly centered, requiring little more than physical clockwork mechanisms to drive them across the sky at just the right rate. The human eye, on the other hand, is physiologically incapable of "integrating" light over more than a fraction of a second. Evolution just didn't select for that sort of vision in our species.

Even though photographic plates inefficiently converted light into a measurable "signal," long-exposure photography using tracking telescopes could vastly increase the effective size of those telescopes' apertures simply by staring at an astronomical object for many hours at a time. Among those with the patience and resources to test its limits, and to absorb the losses due to the inevitable failures of trial and error, photography opened a distinct discovery space that revolutionized astronomy like few technological advances have before or since. Astronomical photographs taken in the second half of the nineteenth century showed that there was seemingly no lower limit to the brightness of stars in our galaxy; revealed the most delicate tracery in the tendrils of gas comprising comets' tails; and led to the serendipitous discovery of previously unknown asteroids and planetary satellites. In the early decades of the new century, photography very nearly displaced traditional observing at the eyepiece in terms of what enabled cutting-edge astronomy and astrophysics.

But why did early twentieth-century astronomers care? The constant push-pull of technology and scientific questioning accelerated exploration to the furthest frontiers well before the dawn of the Space Age. Researchers gradually came to be as interested in the universe beyond the Solar System as they were in the census of objects within it, if not more so. Telescopes weren't then giving up many new secrets about the Moon and planets, but they were revealing astonishing things about stars, galaxies, and even the entire universe itself. That required ever-bigger glass mirrors, more finely controllable telescopes, and increasingly sensitive photographic emulsions, all of which scientists pushed to their limits. The cultural cliché of the lone astronomer seated at the eyepiece in a dark, drafty and often cold telescope dome was only slightly modified: now, instead of recording sketches and impressions in a notebook, the off-axis view through the eyepiece was taken specifically to help guide the telescope's position and keep the photons raining down steadily on the same piece of photographic real estate.

Their patience was rewarded with fundamental discoveries, from the expansion of the universe to the physical processes by which the stars shine. As astronomers became astrophysicists, their planet-watching colleagues slowly turned into planetary scientists who saw connections between geologic structures and atmospheric processes on Earth and their equivalents on the Moon and planets. But while the astrophysicists could simply scale up their machines for detecting and sorting signals from the faint light of the universe, planetary scientists achieved few meaningful gains in terms of their capabilities after the introduction of astronomical photography. Tools such as spectroscopy and observations in other energy regimes of light like the infrared and radio regions of the electromagnetic spectrum added bits of information about the distant places they saw visually through their telescopes; however, meaningful advances would not come until they could send robotic eyes to those places, giving us true bird's-eye views of their physical natures.

Visual observations of planets among professional astronomers declined in importance during these years, with episodes like the Martian canals controversy (Chap. 1) not helping the situation. Figure 6.1 shows the number of scientific papers about the (major) planets published by decade between the mid-nineteenth and mid-twentieth centuries; it suggests that planetary astronomy peaked in popularity in the late Victorian era. Although it didn't fall completely out of favor, there were progressively fewer trained visual observers engaged in planetary work.

At the same time, amateur astronomy was slowly gaining in popularity on both sides of the Atlantic. Although it suffered setbacks during the two World Wars and the Great Depression, public interest in astronomy skyrocketed in the years after 1945. The postwar economic engine put more money in workers' pockets and a combination of advances in labor law and manufacturing technology gave them more leisure time. The affordability of telescopes increased, influenced by the first mass production of suitable instruments, while some people took up amateur telescope making as a hobby unto itself. By the 1950s, vastly more eyeballs were turned skyward, and as Fig. 2.7 shows, Venus observers started reporting Ashen Light sightings in rapidly increasing numbers.

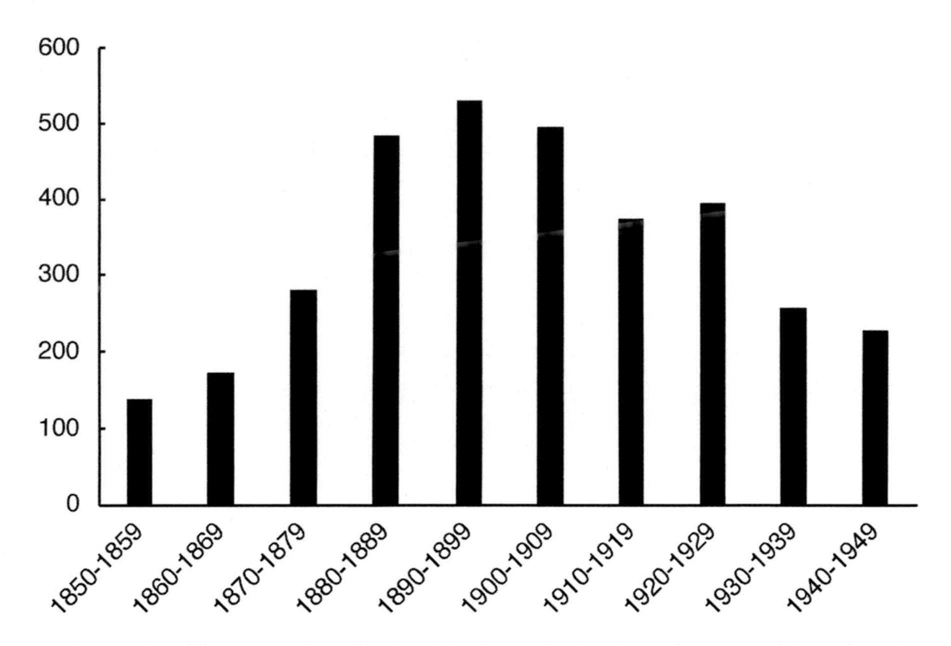

Fig. 6.1 Scientific papers listed in NASA's Astrophysics Data System published during the century between 1850 and 1949 whose titles contained the name of a major planet and whose text contained the word "telescope"

Daylight Observations

As the decades passed, there were more reports of odd appearances of the Venus disc when observed in full daylight, especially around inferior conjunction. C. S. Saxton, an English amateur astronomer in Frome, Somerset, reported an apparent observation of the Ashen Light on the day of an inferior conjunction, September 10, 1927, with an "$8\frac{1}{2}$-inch reflector, [and] a Huyghenian eyepiece, power 132, but the aperture was stopped down to 4 inches in order to obtain a dark field of view." He saw the crescent Venus with horns extending nearly 200° around the circumference of the disk under an exceptionally clear daytime sky. Then:

> when scrutinizing the horns in an attempt to determine their full extent as exactly as possible, the observer was very much astonished at being able to trace faintly the outline of the entire circular disc against the dark blue background. There was no internal edge to the circular outline of the dark portion of the disc, but the whole of this portion showed against the field as a slightly whiter shade of blue.

> The circular outline was very sharp, and the illuminated crescent was irradiating outside the outline of the complete circle."[7]

The explanation for what he saw, Saxton opined,

> must lie either in the observer's eye, or in his telescope, or else in some actual illumination of the dark portion of the disc. Any such illumination, sufficient to make it visible in full daylight, seems somewhat incredible, and it is believed that no one has hitherto been able to account for it satisfactorily, although there are a considerable number of previous observations recorded. … The writer will feel no reluctance in accepting an explanation which will account for the observation by tracing it to the deceit of that very imperfect organ, the observer's eye, or to possible defects in the instrument used.

Others saw discolorations of the Venus disc that rendered it apparently *darker* then the surrounding daytime sky (Fig. 2.9). The French astronomer Roger Cheveau related such an account in what he saw on the afternoon of July 4, 1932, while observing Venus at the Observatory of the Astronomical Society of France in Juvisy-sur-Orge. It was, he declared, "a rather rare observation of this mysterious phenomenon," made in daylight just 5 days after inferior conjunction. "The ashen light is clearly visible," he stated.

> It seems to have a slate-gray, slightly bluish color. Curious mottling, moreover very imprecise, can be distinguished. The most obvious of these differences of hue is constituted by a sort of dark broad band, very faded, substantially parallel to the terminator, to which it is connected by quite distinct kinds of bridges. On the edge of this curious formation, one can sense on the east a very delicate whiteness, and on the left, another clear band, just as vague, but more elongated.

While a sense of scientific skepticism prevented him from drawing any firm conclusions about what he saw, Cheveau's experience offered only "the interest of a rare observation, which makes it valuable to us."

Writing in 2006, Richard Baum recounted the extensive Venus observations recorded by Mark Bernard Brewer Heath (1889–1963), whom Patrick Moore noted in his obituary "ranked as an amateur astronomer" but whose

[7]1927, *Journal of the British Astronomical Assocation*, 38, 64–66. His Scottish colleague, Henry McEwen, made a brief mention of the observation earlier in 1927 in *JBAA*, 37, 345, in which he wrote that Saxton "saw the whole disc, not as a ring, but as a very faint pearly whiteness or a suspicion of mistiness against the blue sky."

"observational work was highly regarded in professional circles."[8] Heath was, as another posthumous tribute had it, "one of those sturdy astronomers who de-iced the dome to get an early morning view of Venus."[9] Baum focused on a "notable series of observations of the dark side of Venus" made by Heath between 1921 and 1953 using the 23-cm reflector at his private observatory in Kingsteignton, Devon, which was in Baum's estimation "undoubtedly one of the most extensive and consistent studies of the phenomenon on record" and "one of the most detailed records of the phenomenon ever kept by a single observer." The record of Heath's observations, comprising individual sightings on 142 dates over three decades, was preserved and published due to Baum's personal interest in the collection of observations.

In nearly every case, Heath described what we would call the "negative" form of the Ashen Light, in which the unlit side of the planet is apparently darker than the background when Venus is observed against the daytime sky. Baum carried out a correspondence with Heath over several years during the 1950s in which Heath was adamant about what he had observed:

> I ought to say that, the pundits notwithstanding, I have seen it [the night side] under so many different conditions, that a sceptic at first, I now have no reasonable doubt of its objectivity. I have seen it both with and without hiding the bright part of the planet, through thick fog, haze, thin cloud, and a lightly tinted neutral Venus glass. It was not always seen on intervening days — or at least was not entered unless visibility was reasonably certain.[10]

But in some instances when Heath followed Venus across the sky past sunset during an eastern (evening) elongation, he saw the same night side appear brighter than the background sky, suggesting a connection between the negative and "positive" forms. Baum again quoted Heath, with emphasis in the original:

> When seen in daylight the unilluminated portion of the disk was invariably noted as darker than the surrounding sky, the darkness being frequently more prominent near the terminator and gradually shading off into invisibility at some point near the line joining the cusps but sometimes extending over the whole or nearly the whole disk. After sunset, at some rather indefinite time (which Mr Baum suggests is about the end of Civil twilight), the dark side has been noted

[8] 1964, *Quarterly Journal of the Royal Astronomical Society*, 5, 295.

[9] Alexander, A. F. 1964, *Journal of the British Astronomical Association*, 74, 122–124.

[10] Heath to Baum in a letter of November 27, 1953, quoted in Baum [4], 191.

> as *brighter* than the outside sky, sometimes showing a dull red or brownish tint and usually this dim glow is seen over nearly the whole of the unlit part of the planet.

A particular example of this effect was noted by Heath on the evening of March 17, 1953, an occasion on which Baum independently reported the "positive" Ashen Light after dusk from his observatory at Chester, some 300 km north of Kingsteignton. At 17:34 Universal Time (UT), some 48 minutes before sunset, Heath saw the night side of Venus as "certainly darker" than the background sky and "dark gray" in color; 1 minute after sunset, he recorded the appearance as only "slightly darker" than the sky and its color merely "gray." But an hour after sunset, with the planet seen against a considerably darker sky background, the night side had become "slightly brighter" than the background, its color "brownish" and texture "mottled."

Writing a decade later, Baum recounted his own recollection of that night, which he found "well worth publishing for its own sake."[11] Observing between 19:00 and 19:40 UT, by which point the planet was "too low for effective observation," Baum noted that the otherwise unlit side of Venus

> was seen to be brighter than the sky, as is the case when the planet is projected against a dark field, and to be veritably glowing with a strong phosphorescent glow of a deep ruddy hue. Upon closer inspection the whole of the dark hemisphere was found to be not uniform in hue but actually mottled over with minute nebulous granules. These were seen better by averted vision.

Baum's perception of the night side of Venus "agreed in every detail with Heath, including the mottled effect and colour." Heath wrote to Baum the following month: "Now I did not know what you were seeing and you did not know what I was seeing. So our observations are absolutely independent, and points of agreement all strengthened thereby." It was this instance, in fact, that caused Heath "to respond positively when asked to list his observations" by Baum, forming the basis for Baum's paper a half-century later.

The American amateur astronomer Walter Haas (1917–2015) was a Venus observer of many decades whose accomplishments included founding the Association of Lunar and Planetary Observers (ALPO) in 1947 and starting its journal, *The Strolling Astronomer*. "Sometimes astronomy advances thanks to group efforts and sometimes due to the perseverance of a single individual. In the passing of Walter Haas ... we have lost someone who excelled at both,"

[11] 1956, *The Strolling Astronomer*, 10, 30–32.

wrote J. Kelly Beatty in his obituary of Haas for *Sky & Telescope*.[12] "At a time when professional astronomers held little regard for amateur observers beyond their meteor and variable-star reports, Haas changed the paradigm."

Based on both his own observations of the Ashen Light and a thorough search of the literature, in 1950 Haas published a short Spanish-language monograph in a Mexican journal called *Urania*.[13] While summarizing the history of sightings of the Ashen Light to date, Haas focused on his own observations and those of others during the period 1948–50. "During December 1949 and January 1950," Haas wrote, "the dark hemisphere was repeatedly seen by several members of the ALPO, so often, in fact, that it does not seem possible to doubt the objective reality of this phenomenon."

Haas mentioned in passing the existence of a photograph of Venus taken by Aden Meinel in 1940 that might have recorded evidence of the "negative" form of the Ashen Light. Meinel (1922–2011), later a Distinguished Scientist at the Jet Propulsion Laboratory and Professor Emeritus at the University of Arizona College of Optical Sciences, was then an 18-year-old technician in the optical shop of Mount Wilson Observatory near Pasadena, California. Around the time of the inferior conjunction of Venus on June 26, Meinel was photographing the planet routinely with a 15-centimeter reflecting telescope to whose optical system a lens was added to deliberately increase its focal ratio. One photograph reportedly showed the region within the crescent as darker than the surrounding sky, but it has evidently disappeared from the historical record.[14] Fifty years later, Richard Baum commented that "It is a matter of some regret that so far as is known this observation was not published ... in spite of efforts to trace the photograph, ... [it] is now presumed lost."

Haas also argued in favor of the reality of the "negative" Ashen Light in opposition to commentators such as Patrick Moore, who believed it was a trick played on the brain by a contrast effect. Haas confronted such skepticism directly: "Although it has been assumed that the appearance of the dark hemisphere is an optical illusion because it is not perceived in full darkness,

[12]"Walter H. Haas (1917–2015)" (April 8, 2015) https://www.skyandtelescope.com/astronomy-news/walter-haas-1917-2015-04072015/.

[13]1950, La Visibilidad del Hemisferio Oscuro de Venus, *Urania*, 223–224.

[14]Haas attributed the tale to "Mr. C.B. Stephenson, a graduate student at the Yerkes Observatory, and director of the Mercury Section of the ALPO" from a personal letter "in which he describes that Dr. A.B. Meinel has successfully achieved photographs within the narrow crescent of Venus, which was darker than the surrounding sky." Stephenson specifically noted in his communication that Meinel's photograph was taken "with ultraviolet light." This probably involved use of a near-ultraviolet filter passing wavelengths a little shorter than 400 nanometers; below about 300 nanometers the Earth's atmosphere becomes completely opaque to ultraviolet light due to absorption by ozone (O_3) molecules. To date, it appears there is no other known data on this phenomenon in the ultraviolet region of the spectrum.

it should be noted that the observers of the ALPO noticed this phenomenon when the Sun was over the horizon, during a moment before or after sunset, and when the sky had almost darkened."

So if the Ashen Light is something real and inherent to Venus, Haas wondered, "what is its explanation?"

> We can not seriously assume that it is a mere reflection. The Earth is the brightest star, except the Sun, in the Venusian sky. Since the Earth does not illuminate the lunar surface so that we can see it in the daytime sky, the much fainter light of the Earth at the greatest distance from Venus cannot produce a greater effect. Noble and others suggested that the visibility of the dark hemisphere is due to the silhouette of the planet projected against a bright background produced by the zodiacal light or the outer solar corona.[15] This explanation demands that the [unilluminated] hemisphere of Venus be darker than the sky. Dr. Meinel considers this effect as the one he photographed and it can scarcely be doubted that he also explains many visual observations made near inferior conjunction when the dark hemisphere is seen as more intense than the background of the sky.

Still, Haas conceded, "it would be very curious [if] the [sky] background [behind] Venus was always the external solar corona or the zodiacal light." He considered both explanations "doubtful," in part because a silhouetting effect against a bright background external to both Venus and our own atmosphere wouldn't explain why some observers saw texture or some other kind of inhomogeneity in the perceived intensity of the unlit side of the planet.

In the end, Haas entertained an idea to which we'll return later in this chapter. "It is interesting," he wrote, "perhaps too bold, to surmise that the blue tint of the dark hemisphere, mentioned earlier, is related to the green auroral line of 5577 angstroms." He concluded with several specific recommendations for future studies of the Ashen Light:

(i) Attempts to photograph the dark hemisphere, even with ultraviolet light.
(ii) Careful photometric measurements, perhaps visual and photographic, of the brightness of the dark hemisphere and the sky next to it.
(iii) Attempts to determine the color of the dark hemisphere, perhaps by means of photography with color filters.

[15]Haas here quotes Thomas William Webb, 1917, *Celestial Objects for Common Telescopes* (Sixth Edition), Vol. I, 74–76. "Noble" is Captain William Noble (1828–1905), an officer in the British Army who later retired to Sussex and built a private observatory, contributing to several English astronomical organizations over a half-century.

(iv) Special examinations of the dark hemisphere when there is great [magnetic] activity on the Sun.
(v) Very careful examinations of the dark hemisphere during the rare opportunities provided by the total solar eclipses. The corresponding diminution of brightness of the sky near the Sun must allow meaningful and informative observations to be made well.

The Barbier Campaign (1935)

Referring to the pop cultural phenomenon then current in the West, in 1934 a French astronomer named Daniel Barbier (1907–1965) wondered whether "perhaps the Loch Ness Monster has prompted us to consider this kind of 'astronomical sea serpent' that is the ashen light of Venus."[16] While himself skeptical of not only the historical accounts, but also all of the proposed explanations, Barbier made one of the first attempts preserved in the scientific literature to determine whether it is reasonable to believe that the Ashen Light has a physical nature having to do with Venus itself based on the question of whether it should even be visible from the Earth in the first place.

To arrive at an answer, Barbier considered the circumstances of historical reports, noting that "the diameter of the instruments that have been used varies from 45 to 760 mm, but only occasionally exceeds 200 mm." So whatever it might be, it seems biased toward detection through smaller apertures—but only to a point. On that basis, he worked out how bright diffuse light from an extended object must be in order for the eye to stand a good chance of detecting it through a small refracting telescope on Earth. As his threshold of detection, Barbier assumed 10% of the background; with a literature figure in hand for the brightness of the daytime sky, he could then assign a numerical value for an "observable" Ashen Light under conditions of full daylight:

> The brightness of the diurnal sky at 10° from the Sun in a very clear sky is of the order of one candela per square centimeter.[17] In order for Venus' disk to be seen apart from the [background] sky, it can be assumed that its brightness should be about 1/10 of that of the sky, because we must take into account the causes of trouble brought by the secondary spectrum of the objective and the diffusion of light by the glasses of the instrument.[18] We will therefore admit that

[16] 1934, La Lumière Cendrée de Vénus. *L'Astronomie*, 48, 289.

[17] Citing J. Payer, 1927, *Revue d'Optique*, 6, 73.

[18] "Secondary spectrum" refers to the colored fringes around objects induced by poor chromatic aberration correction in the objective lenses of refracting telescopes.

> the brilliance of the ashen light of Venus is 0.1 candela per square centimeter or, what amounts to the same, of −12.9 magnitudes per square degree.[19]

Citing the surface brightness of the full moon as 0.25 candela per square centimeter and the Earthshine as 0.0006 in the same units, Barbier concluded that "the brightness to be explained is much greater than that of the ashen light of the Moon [Earthshine] and almost equal to that of the Moon itself."[20] This result is not a surprise, given the familiar experience of seeing the directly illuminated part of the Moon against the daytime sky. While not accounting for other details, like the angular size of the object, the surface brightness of the Moon is a first-order guess for how bright something must be in order for an observer to clearly see it during the daytime. Keen observers can see both Venus (∼3 candela per square centimeter) and Jupiter (∼0.08 candela per square meter) during the daytime.

According to Barbier's reckoning, Ashen Light seen during the daytime—that is, the night side of the disc seen as *brighter* than the surrounding sky—would be about 33 times fainter than the sunlit portion of the disc. Bear in mind, however, that this estimate does not consider the circumstances under which the Ashen Light is mostly reported: a twilight sky background, or even full ('astronomical') darkness. The threshold observability of the Ashen Light would scale down in brightness proportionally under lower ambient light levels.

He then compared his number to surface brightnesses predicted by various hypotheses for the cause of the Ashen Light. First, he eliminated projection against the zodiacal light or outer solar corona in order to cause the Venus night side to be seen in silhouette; 40° from the Sun, around where Venus reaches greatest eastern or western elongation, the zodiacal light "would be

[19] Payer, 293.

[20] Modern values for these quantities vary from those Barbier assumed. In 2017, Christopher Kyba, Andrej Mohar, and Thomas Posch (How bright is moonlight? *Astronomy & Geophysics*, 58(1), 1.31–1.32) quoted a value of the maximum possible horizontal photopic illuminance due to full moonlight as ∼0.3 lux, or about 33,000 times less intense than direct sunlight. To find the corresponding luminance, or surface brightness, of the Moon, we need to know the solid angle it subtends. On average, the Moon has an angular diameter of about 31 minutes of arc, corresponding to a solid angle of $\sim 6.4 \times 10^{-5}$ steradian. This implies that the brightest possible full Moon has a luminance of about 5000 candelas per square meter, which is roughly twice as large as Barbier's value. Meanwhile, the Earthshine is observed to have surface brightnesses ranging from 13 to 16 magnitudes per square arcsecond, or 0.04–0.7 candela per square meter depending on the lunar phase angle and the Earth's weather-dependent albedo (Montañés-Rodríguez, P., Pallé, E., & Goode, P. R. 2007. Measurements of the Surface Brightness of the Earthshine with Applications to Calibrate Lunar Flashes. *The Astronomical Journal*, 134(3), 1145–1149). This is anywhere from 8 to 150 times smaller than Barbier's number. Given the modern values of the luminance of the full Moon and the Earthshine, their ratio is on the order of 7,000 to 125,000.

3.6 magnitudes per square degree, or about 25×10^{-9} candela per square centimeter, or about 10 million times smaller than the brilliance that we would need."[21] The solar corona fails the test as well, since it "decreases very quickly in intensity when one moves away from the Sun and its brilliance within 2° of it is practically nothing."[22]

Barbier then tried direct illumination of Venus by Earth near inferior conjunction, finding that the required illumination level is over a hundred times higher than the Earthshine for an object orders of magnitude further away than the Moon. Since the calculated surface brightness due to reflected Earth light at the cloud tops of Venus is only about as bright as the zodiacal light in the previous situation, Barbier concluded, "the hypothesis is therefore abandoned." In any case, he points out, the increase in the illumination of the night side of Venus from Earth is almost exactly canceled out by the background attributable to the zodiacal light against which Venus is seen, effectively nullifying any possible contrast between them.

Next, Barbier considered planetshine from a hypothetical Venus satellite. Correctly concluding that it "should have a low brightness, since despite all the research it has never been seen," he computed that in order to make this mechanism work to explain the Ashen Light, a putative moon would have to be nearly as bright as Venus itself. Depending on the mean distance between Venus and its moon, Barbier calculated corresponding equivalent stellar magnitudes as high as −10, far brighter than the planet itself. In addition, he discounted the possibility that it could be much closer to Venus, and hence at a much lower brightness, because it would be deep within the shadow of the planet at that point and not illuminated by the Sun at all. "It is," Barbier wrote, "not useful to dwell more on the absurdity of this result."

---

[21]This result is reasonably close to a modern value. Leinert et al. (1998. The 1997 reference of diffuse night sky brightness. *Astronomy and Astrophysics Supplement Series*, 127(1), 1–99) cite a brightness at 500 nm for the zodiacal light 40° from the Sun along the ecliptic of 925 in a unit called S_{10}, where one such unit equals the brightness equivalent to that of one tenth-magnitude star of solar spectral type per square degree of sky. Given that the zodiacal light is scattered sunlight, it has a spectrum like that of the Sun and an effective temperature of 5800 kelvins; using the temperature-dependent conversion between magnitudes per square arcsecond and luminance given by Bará et al. (2020. Magnitude to luminance conversions and visual brightness of the night sky. *Monthly Notices of the Royal Astronomical Society*, 493(2), 2429–2437), the zodiacal light at this position on the sky has a surface brightness of about 9.7×10^{-8} candela per square centimeter, or a factor of four higher than Barbier's value.

[22]Aggregating literature data published between 1955 and 1996, Kimura and Mann (1998. Brightness of the solar F-corona. *Earth, Planets and Space*, 50(6–7), 493–499) found that the surface brightness of the solar corona 2° from the Sun in its equatorial plane is about 4×10^{-10} that of the surface brightness of the Sun itself. This is still some 5,000 times higher than the surface brightness of the night sky, so it's arguable whether Barbier's characterization of the corona at this elongation as "practically nothing" is reasonable.

After evaluating and tossing several other more obscure ideas, Barbier tested one more: whether the Ashen Light was something like the Earth's aurorae. At their brightest, terrestrial aurorae do not exceed around 10^{-3} candela per square centimeter, which is about one hundred times weaker than Barbier's expectation for the brightness of the Ashen Light. Since in the 1930s it was not unreasonable to assume that Venus must be possessed of an intrinsic magnetic field like Earth's, Barbier did not think that "this hypothesis is to be rejected absolutely." He thought that an auroral explanation probably accounted for the sometimes apparently inhomogeneous Ashen Light, which "is not evenly distributed on the disc, but centered on the pole of the dark hemisphere, which would be a magnetic pole."

He concluded that if the Ashen Light is real, it is likely an auroral phenomenon:

> It is extremely probable that the greatest number of observations of the ashen light is of purely subjective origin, the illusion being facilitated by atmospheric or instrumental phenomena or sometimes even by twilight phenomena on Venus. The only cause by which one could explain the supposedly real ashen light would be the presence of polar auroras at least a hundred times more intense than on the Earth.

Barbier also suggested collaboration among observers as key to understanding whether the Ashen Light is real or not. In order to answer that question, he advocated for what appears to be the first coordinated Ashen Light observing campaign in history. "The study of these appearances," he wrote in 1935, "seemed to us to require the collaboration of a large number of observers working simultaneously and with identical processes, which is why we called upon the collaboration of available astronomers."[23] With a favorable inferior conjunction coming later in the year, Barbier hoped to recruit the participation of many astronomers to the effort. Observations contributed by "the rare owners of reflective telescopes" were particularly desired, since they didn't suffer the colored haloes attendant to chromatic aberration in refractors.

Advice given to participants included taking care to avoid allowing direct sunlight to fall into their apertures; the use of field stops and occulting bars to hide the glare of the illuminated crescent and to employ colored filters cutting the blue of the daytime sky and violet light from chromatic aberration. "For reasons of economy and unification," Barbier threw in, "we ask each observer to be willing to provide a small piece of red photographic lantern glass."

[23]1935, Recherche de la Lumière Cendrée de Vénus. *L'Astronomie*, 49, 264–268.

Observers were asked to respond to a series of questions about each instance where the Ashen Light was suspected:

Was the entire disk of the planet seen?

(a) *Without a colored filter. With low magnification, with medium magnification. By hiding the crescent. If the disc is seen, does the appearance change by slightly changing the focus?*
(b) *With a colored filter. Same questions except the last one that becomes irrelevant.*

In the case where one of the above observations would have shown the disc, the following questions should be answered:

Is the disc lighter or darker than the sky background?
Does it appear uniform or degraded? Is its extent clear?
What is its color (without a colored filter, of course)?

Twenty-one individuals heeded Barbier's call, sending him 213 observations made between July 1 and October 7, 1935.[24] Most observers did not report seeing the Ashen Light during the campaign. Those that did report something tended to see a bluish darkening either inside or outside of the crescent. Barbier ascribed this to either chromatic aberration where refracting telescopes were used or a contrast effect "involving persistence of the image of the bright crescent on the retina, which manifests (through bleaching)[25] as a dark after-image."[26]

"In summary," Barbier wrote with emphasis in the original,

to explain the observations made during this conjunction, we see that there is no need to involve causes arising on the disc itself. Virtually all the observations made during previous conjunctions happen to have their equivalent in the

[24] Barbier, D., & Schlumberger, R. 1936. Recherche de la Lumière Cendrée de Vénus Pendant la Conjonction Inférieure de 1935. *L'Astronomie*, 50, 27–34.

[25] "Bleaching" here refers to the destruction of cone opsins, light-sensitive proteins in the retinal cone cells of the eye. At high, persistent light levels, bleaching of photopigments in the cone cells leads to a decrease in sensitivity to light while the photopigments regenerate. As Barbier suggests by his comment, the persistent, focused light of the Venus crescent as seen through the eyepiece of a telescope can create after-images on the retina if the eye position suddenly changes after several seconds of continuous observing. He thought this must explain why observers saw dark shapes adjacent to the crescent under such circumstances, which in this hypothesis result from a photochemical process in the retina and not in the atmosphere of Venus.

[26] Barbier's co-author, Rene Schlumberger, in a separate commentary (pp. 32–34), dismissed chromatic aberration as a cause of supposed Ashen Light observations and proposed the contrast effect as an alternative, noting that "...observing Vega, by day, our instrument has never shown us the violet halo, due to the secondary spectrum, while at night it is quite apparent."

observations we have just analyzed, and *it can be concluded that the dark part of the disc of Venus is really invisible to us and that the ashen light is a legend.*

Nikolai Kozyrev and the Venus Airglow

Nikolai Alexandrovich Kozyrev (1908–1983) is remembered by history for his controversial observations of light emissions from the Moon during the 1950s that were subsequently claimed as evidence for contemporary geologic activity on the Earth's seemingly dead nearest neighbor. His observations of Venus, although maybe of more lasting significance, are usually relegated to a footnote. But they may hold the key to understanding the Ashen Light.

Kozyrev was born in St. Petersburg, then the Imperial Russian capital, and received a degree in astrophysics from Leningrad State University at age 20. He moved on to nearby Pulkovo Observatory in 1931 and seemed to have a promising scientific career ahead of him. However, during the 1930s he became caught up in the sweeping Stalinist purges of many Soviet academic institutions. In November 1936, Kozyrev and several of his colleagues at Pulkovo were arrested on suspicion of seditious activities raised by the scurrilous allegations of a disgruntled former graduate student who worked at the observatory. Sentenced to 10 years in prison for "counterrevolutionary activities," Kozyrev fared better than many of his co-workers who were executed or later died in the gulags.

1941 saw another decade-long prison term imposed on him for spreading "hostile propaganda." His supporters on the outside lobbied for his release, which was granted early in January 1947. His diligent pursuit of science under hostile conditions won him a mention in Alexandr Solzhenitsyn's *The Gulag Archipelago*, in part for which Solzhenitsyn was awarded the 1970 Nobel Prize in Literature. Upon his release, Kozyrev went back to work and tried to rebuild a career shattered by the political machinations of the Soviet state. He wound up again at Pulkovo, his academic home for the rest of his working life.

In the autumn of 1958, Kozyrev was involved in a program of observing lunar craters with a spectrograph attached to the Crimean Astrophysical Observatory 1.3-meter telescope. The instrument dispersed the light collected by the telescope into its component colors, allowing the user to search for the chemical fingerprints of different elements upon inspection of photographic plates that recorded the object's spectrum. During an exposure on the central peak in the crater Alphonsus, Kozyrev claimed to have seen a brightening of the area around the mountain. After developing his photographic plates, he further claimed the detection of features in the spectra he attributed to a rarefied cloud of diatomic carbon molecules. The mountain must have emitted the cloud of gas, he hypothesized, which was then caused to fluoresce by exposure to ultraviolet light from the Sun. While other observers had previously noted fleeting bright spots on the lunar surface, no one before Kozyrev purported to have evidence explaining what they were. While the question of whether the Moon might yet be geologically active lacks a definitive answer, Kozyrev's methods and conclusions remain more controversial still.

It's with an appropriately sized grain of salt, therefore, that we should take the case of another remarkable spectrum that Kozyrev obtained several years earlier. In March 1953 he made several exposures on the night side of Venus, again with the CAO telescope and spectrograph. Here, too, he came up with an unexpected result: optical light emission in the Venusian night. He found three spectral lines on his plates that matched the known wavelengths of energy transitions of ionized nitrogen molecules (N_2^+).

The explanation for these lines, Kozyrev indicated, was an analog to "airglow" on Earth that we encountered briefly in Chap. 5. Airglow (Fig. 6.2) is a well-understood phenomenon of the terrestrial atmosphere that results largely from the excitation of its constituent gases to highly energetic states. It is perceptible to dark-adapted observers on the ground, and sometimes shows up in photos of the night sky as a faint, rippled glow stretching across all or part of the sky. Although it varies in intensity and sky distribution over durations as short as a few minutes, airglow is more or less omnipresent in the atmosphere and results in a natural limit to how dark the night sky can ever be. Kozyrev was certainly well-acquainted with the phenomenon, and it was no stretch of the imagination to imagine an analogous process might be at work in the atmosphere of Venus.

Five years later, after Kozyrev published his Venus results and seeking to repeat the observation, Gordon Newkirk (1928–1985) of the University of Colorado made similar measurements that seemed to indicate the same features Kozyrev identified. In addition to confirming two of Kozyrev's nitrogen lines, at wavelengths of 441 and 443 nanometers (nm), Newkirk found a third nitrogen line at 450 nm that his Russian colleague evidently missed. He calculated that the intensity of the 443-nm line implied light emission about 80 times higher than that of terrestrial airglow, which should have rendered it easily visible through the telescope eyepiece on Earth.

Yet Newkirk didn't detect it visually. Over the several nights of these observations in January 1958, he wrote,

> I carefully examined the dark side of Venus for emission by allowing the bright crescent to pass through the slit aperture and reflecting the dark side from the

Fig. 6.2 Airglow seen over the limb of the Earth from aboard the International Space Station. Different physical processes yield the colorful layers that form at altitudes between 90 and 300 km. The colors indicate the elements emitting light: yellow-orange from sodium atoms, and green and red from different excited states of oxygen. NASA photo ISS042-E-037847 taken December 15, 2014, by European Space Agency astronaut Samantha Cristoforetti

polished slit jaw. On no occasion was any radiation from the dark side visible (although Patrick Moore of the British Astronomical Association reports that the ashen light was visible under high (×320) magnification during this period).

The whole episode mirrored the history of Ashen Light observations generally, in which the underlying mechanism was some sort of on-again, off-again affair. That the two observers saw different things on different nights was not inconsistent with expectations. "If the glow is an auroral type," the Dominion Astrophysical Observatory's Andrew McKellar wrote in 1960, "it would be expected to vary in brightness from night to night, just as do our northern lights." Here was a solution to the inconsistently bright spectral features on the night side of Venus. "Therefore," McKellar argued, "it is possible that the results of Kozyrev and Newkirk can be reconciled."[27]

The calculated brightness of the nitrogen emission in Newkirk's spectra was comparable to that of aurorae in the Earth's atmosphere, which are often easily visible to the unaided eye. Newkirk argued that even so, under typical conditions, the spectral line emission was just too weak to sufficiently stimulate the human retina to detection:

> A source with an apparent diameter of 30′ (as seen in the eyepiece of our apparatus) with a radiance of 20,000 R [Rayleighs][28] is only a factor of about 40 above the threshold of the eye for peripheral vision. Considering the fact that the eye is some five times less sensitive at 4300 Å than at 5577 Å we should not be surprised that the ashen light, which is superposed on a field of scattered light, is difficult to see. An increase in the intensity of the band by a factor of five would bring the ashen light well above the threshold of visibility. Such a phenomenon may account for the occasional sightings of the ashen light.

Newkirk suggested no particular means by which nitrogen molecules could be stimulated to yield sufficiently strong emission to ensure confident visual detection, but he made a useful remark about the nature of "peripheral" vision. This refers to a particular biological route for sensing light in the human eye. Astronomers have known for centuries that looking at a faint celestial object through a telescope eyepiece just slightly off-center causes the object to suddenly appear brighter; in fact, for a celestial object at the threshold of visual detection, "averted" vision may well mean the difference between seeing it or not seeing it.

The reason has to do with the concentration of rod cells in the human retina, which peaks away from the fovea, a small region near the center of the field of view that is tightly packed with color-sensing, but relatively light-insensitive, cone cells. The rods give exquisite sensitivity to faint light but offer no color information to the brain. Therefore, Newkirk implies a value of 500 Rayleighs for the threshold of detection by the human eye specifically under conditions

---

[27]1960, *Journal of the Royal Astronomical Society of Canada*, 54, 99.

[28]A unit of surface brightness, the Rayleigh is defined as follows: one Rayleigh corresponds to a rate of light emission of 10^6 photons per square centimeter of column per second. A 'column' in this context is integrated along the line of sight such that a three-dimensional object, like a cloud of emitting material, is collapsed down into a two-dimensional cross section—hence "per square centimeter." This accounts for the fact that we view three-dimensional light emitters, assuming they remain transparent to radiation throughout, in projection.

Table 6.1 Surface brightness of various phenomena expressed in units of Rayleighs. Data sources include Goody and McCord [24]; Cruikshank, Hartmann and Landman (unpublished data, 1979); Cox [16]; and Napier [49]

Source	Brightness (Rayleighs)
Daytime side of Venus	9×10^{10}
Daytime sky on Earth	9×10^{9}
Night side of Venus	5×10^{8}
Very bright (class IV) aurora	2×10^{6}
Cumulus clouds illuminated by full moonlight	10^{5}
Faint aurora	10^{4}
Milky Way near the Galactic equator	1300
Night sky near zenith	1600
Night sky near horizon	1000
Nighttime airglow near zenith	800
Night side of Venus illuminated by Earth and starlight	500
Zodiacal light away from the ecliptic	100

of averted vision. To convey an idea of how bright this is, Table 6.1 gives surface brightnesses for a variety of night sky phenomena in Rayleighs.

Kozyrev's work in the 1950s raised the possibility that some kind of physical or chemical conditions in the atmosphere of Venus might give rise to the emission of enough visible light on the planet's night side to render a glow visible with Earth-bound telescopes. Early spacecraft missions to Venus in the 1960s revealed that the planet was emitting much more ultraviolet light than expected and that at least some of that light originated on the night side. *Mariner 5* spacecraft results suggested that some kind of atmospheric chemical reaction could explain the emission, or possibly the bombardment of the atmosphere by charged particles from the Sun. Seven years later, *Mariner 10* flew past Venus and again detected extreme ultraviolet emissions ten times brighter than theorists predicted, but it remains unclear how this emission, otherwise certainly invisible to the human eye, might be perceived as something like the Ashen Light.

The International Geophysical Year (1957–1958)

Ashen Light reports again spiked during the favorable evening apparition of Venus in the winter of 1957–1958, culminating in the inferior conjunction of 28 January. This episode was situated neatly within the International

Geophysical Year (IGY), a scientific collaboration among some 67 countries lasting from July 1957 to the end of 1958. The IGY corresponded by design with the expected peak of Solar Cycle 19, which turned out to have the most intense maximum of any solar cycle since the late eighteenth century. Global cooperation during the IGY focused on 11 Earth science topics ranging from geomagnetism and upper atmospheric physics to oceanography and seismology.

Around the same time, the discipline of planetary science as we now know it was on the eve of its birth as the Space Race began. The world's first artificial satellite, *Sputnik 1*, was launched into orbit by the Soviet Union in October 1957. As spacecraft were trained on the Earth's Moon and other planets, planetary science blossomed as a holistic scientific approach to understanding other planets as worlds unto themselves, and tools such as geomorphology, atmospheric science, and geochemistry were brought to bear in the analysis of data obtained directly from on and near planetary surfaces for the first time. In this, it expanded beyond what would have previously been characterized as planetary astronomy, in which planets were treated as distant objects whose study was only made accessible through telescopic observations from the Earth.

Among the founders of modern planetary science was Gerard Kuiper (1905–1973), a Dutch-American astronomer who transformed planetary studies during his decades-long career at the University of Chicago. In 1960 he moved west to found the Lunar and Planetary Laboratory at the University of Arizona in Tucson. Among his early graduate students there were Dale Cruikshank and Alan Binder, who had spent their undergraduate summers as research assistants at the University of Chicago's Yerkes Observatory in Williams Bay, Wisconsin.

Cruikshank's career-long interest in Venus took off during this time, when he had at his disposal for planetary observations what remains the world's largest refracting telescope, the 1.02-meter Great Yerkes Refractor. The telescope was truly a beast, with its 19-meter-long tube sometimes only reached by the full six-meter extension of an elevator floor that lifted the observing platform up to meet its eyepiece. The Yerkes summer students were offered unrestricted use of the venerable old telescope in the daytime, and they took advantage of that particular liberty to study Mercury and Venus at midday, when looking through the tremulous atmosphere in their direction yielded the least distortion of the planets' images.

Their intent was to search visually for the hazy shades and markings on both planets' faces reported by observers over the centuries. However, as Cruikshank related to me some six decades later, "Al and I reluctantly and independently came to the same conclusion that we could catch glimpses of the unilluminated

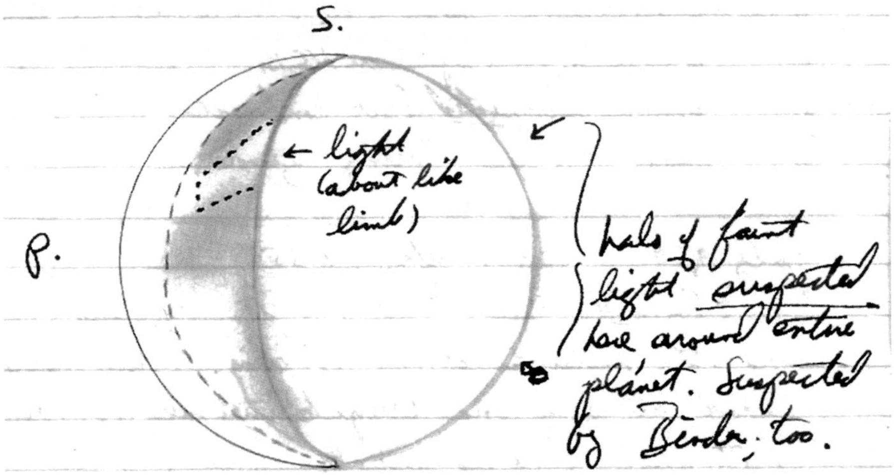

Fig. 6.3 A drawing of Venus made by Dale Cruikshank in his observing notebook entry of July 20, 1959, showing the "halo effect" he and Alan Binder saw: "Halo of faint light *suspected* here, around entire planet. Suspected by [Alan] Binder, too"

part of the Venus disc, defined against the bright background sky by a faint halo encircling it and marking the planet's limb." (Fig. 6.3)

Cruikshank interpreted what he saw as the Ashen Light. His observing notebook of July 11, 1959, records its first appearance, with emphasis added in the original:

> 'Ashen Light' *suspected*, especially around the cusps. This made the dark side appear darker than the surrounding background sky. *Extremely difficult*. ...The Ashen Light was suspected with the planet's dark side appearing slighter darker and of different general color than the background sky.

He added a marginal note on Boxing Day in 1959, when he and Binder were briefly back at Yerkes during the semester break:

> The above description of the Ashen Light was given mostly to describe the halo of faint light as recorded in the following observations of Venus in July.

Recounting his firsthand observations in 1983, Cruikshank wrote "I am convinced the phenomenon is real."[29] And I knew I had to ask him about it.

[29]The development of studies of Venus, *Venus* (ed. D.M. Hunten, L. Colin, T.M. Donahue, and V.I. Moroz) Tucson: University of Arizona Press, 8.

Speaking to me from his office at NASA's Ames Research Center in Mountain View, California, Cruikshank recalled in clear detail what a much younger version of himself perceived decades earlier. "For the next few days," he continued, "Al and I saw the glow defining the night hemisphere off and on, and made notes. We knew that these observations were unusual because the traditional Ashen Light is reported at much higher phase angles than those we were working at in July."

I brought up the historically strong solar activity of the IGY, knowing full well that he was aware of the apparent correlation between Ashen Light reports and the phase of the solar cycle. It was only after he and Binder made their observations, he said, that they learned several strong solar flares were recorded around the same time. They plotted the dates on which they saw what came to be known between them as the "halo effect" as well as the strong solar flares, and a correlation became evident.

The whole thing was set aside as the two enrolled in the graduate program at the nascent University of Arizona Lunar and Planetary Laboratory. But they dusted off the observations in early 1962 and related what they saw in the draft of a scientific paper tentatively titled "Solar Activities and Observations of the Nocturnal Luminescence of Venus." In it they argued for the reality of what they saw, tying its appearance to the strong solar maximum of Solar Cycle 19 under the presumption that the enhanced solar activity was exciting something like an aurora in the Venusian atmosphere. After completing the draft in the spring of 1962, they duly sent it to Gerard Kuiper for his consideration—and, hopefully, his blessing—before subjecting it to formal peer review.

But Kuiper wouldn't have any of it. He wrote a rather formal memorandum to the two students on May 19, noting that he showed the manuscript to Aden Meinel. "It is our opinion," Kuiper wrote very matter-of-factly, "that the reality of the reported observations is extremely doubtful. ... As Dr. Meinel states: We do not want to 'pull another Kozyrev.'" A claim of the nature that Cruikshank and Binder made in their paper was

> bound to attract immediate attention and cause a great deal of discussion. It would be most unwise for the [Lunar and Planetary] Laboratory to make startling statements which within a few months may have to be withdrawn completely. I must, therefore, ask you to hold this paper in reserve and not publish it in any form whatever.

Kuiper wanted confirming data before he would consider allowing them to publish, which Cruikshank told me "of course were never made, or at least never published, and our paper never saw the light of day."

Cruikshank was emphatic that he and Al Binder saw *something* happening on Venus in the summer of 1959. "I still think that Al and I made valid observations," he said confidently. He continued to ponder this problem during the intervening years that saw his own successful career in planetary science, and he remains open to the idea that the Ashen Light is a phenomenon rooted in physical reality. Under unusual circumstances catalyzed by strong solar flares, he reasoned, the atmosphere of Venus might glow in visual wavelengths with sufficient intensity to be detectable to the trained human eye.

The notion of a "halo" around Venus jogged my memory as I listened to Cruikshank recall what was clearly an exciting time to be working in planetary astronomy. Many observers have seen, and even photographed, such an effect in the hours around inferior conjunction. As the illuminated crescent shrinks in width, it begins to elongate in the azimuthal direction running around the circumference of the planet's disk; astronomers have noted these "cusp extensions" nearly since the invention of the telescope (and photographed them; see Fig. 3.4).

This variety of halo effect, however, is not fleeting or controversial in nature. Rather, it is the result of forward scattering of sunlight through the planet's atmosphere. Light from the Sun landing near the edge, or limb, of the planet as seen from Earth can be redirected by interactions with small particles in the Venusian atmosphere. Since the length of the path traversed by light rays is a maximum near the limb, the chance that a particular ray will scatter during its grazing encounter with the planet's atmosphere is relatively high. Yet this can't explain what Cruikshank and Binder thought they saw, observing roughly 40–50 days before inferior conjunction. Maybe it was some other, more exotic form of scattering with a different geometry?

Cruikshank was a firm "no." "If it were not," he reasoned, "wouldn't it imply a characteristic of Venus' atmosphere that's not normal or what we think it is?" I couldn't easily argue the contrary, and definitely not to a person who knows vastly more about planetary atmospheres than I do. By all rights, we should expect to see optical effects in the atmospheres of other planets familiar from our experiences on Earth.

That leaves only one of two explanations to account for all of this: either something real in the atmosphere of Venus (possibly triggered by solar activity), or a purely perceptual phenomenon that exists only in the mind of the observer. "If these really are temporally anomalous events," Cruikshank suggested, "someone has to be out there looking. Interesting things do pop out on a short timescale, and unexpectedly." As for the ultimate reality of observations he made most of a lifetime ago, Cruikshank is ambivalent: "Even though I stand by what I wrote in my notebook, I allow for the possibility that it's all illusory."

Ashen Light sightings continued into the early 1960s just as humanity was about to get its first close-up views of Venus from spacecraft. I remembered reading accounts of some of these observations, finding a familiar name among them: the planetary scientist William K. "Bill" Hartmann, a senior researcher at the Planetary Science Institute in Tucson, Arizona. In the early 1960s, Hartmann was another student of Gerard Kuiper's. I first met him in the mid-1990s when I was an undergraduate student at the University of Arizona taking a seminar course on Mars in the department Kuiper founded.

Hartmann is scientific virtuoso with expertise in many fields of planetary geology, including Mars. He's also a master of photographic techniques who was called to testify before the U.S. House Select Committee on Assassinations in 1978, giving a detailed report on the photographic evidence associated with the 1963 assassination of President John F. Kennedy. During the same decade, with PSI scientist Don Davis, Hartmann hypothesized a scenario in which the Earth's Moon was formed in the aftermath of the impact of a Mars-sized object that has come to be called Theia. The "giant impact hypothesis" is today the leading idea explaining how Earth came to have a moon. On top of it all, Hartmann became an accomplished space artist whose work graces the pages of many popular books about the Solar System.

But he's also a thoroughly experienced visual planetary observer, as I discovered when I visited him in his office at PSI. Rifling through several filing cabinets that contained decades' worth of observing and research notes, preprints of scientific papers, and the other ephemera of a long scientific career, there was the "eureka" moment. "Ah," he said, confidently. "Here we go. Venus." He extracted from the contents a thick file comprised of many dozens of looseleaf pages, some of which represented records of observations he made as a teenager in the 1950s.

Among the papers were a set of notes that he, Dale Cruikshank and Al Binder made in November 1962 before and after that month's inferior conjunction on the 13th. The three men used Cruikshank's 12-inch reflecting telescope for the daytime observations, made all the more challenging by the nature of the telescope's open-truss tube design that allowed light from the daytime sky to scatter through the telescope's optical path (Fig. 6.4). To make the situation more tolerable for the observer, one of the others would hold up a screen to block direct sunlight from falling on the top end of the telescope, where it was most likely to scatter and cause interference with the observation.

Fig. 6.4 Al Binder and Bill Hartmann observing Venus near the Sun with Dale Cruikshank's 12-inch reflecting telescope in Tucson on November 9, 1962. Hartmann (right) holds a shade to prevent direct light from the Sun from falling on the top end of the telescope. Photo by Dale Cruikshank

The three were primed to look for anything out of the ordinary on the night side of the planet in part as a result of Dale and Al's experience at Yerkes a few years earlier. But day after day, they saw nothing, as suggested by excerpts from Hartmann's 1962 observing notebook:

November 7: *No evidence of dark side lighter or darker than sky.*
November 8: *Halo never clearly seen.*
November 10: *No dark side phenomena seen.*

Remarks to the same effect continued as late as November 22, but one date stands out: the 12th. "Ashen Light strongly suspected," Hartmann wrote in his notebook that day (Fig. 6.5). He continued, "A copperish hue brightest on crescent side." Hartmann and Cruikshank agreed to draw what they saw separately before they talked about it so as not to influence the other's

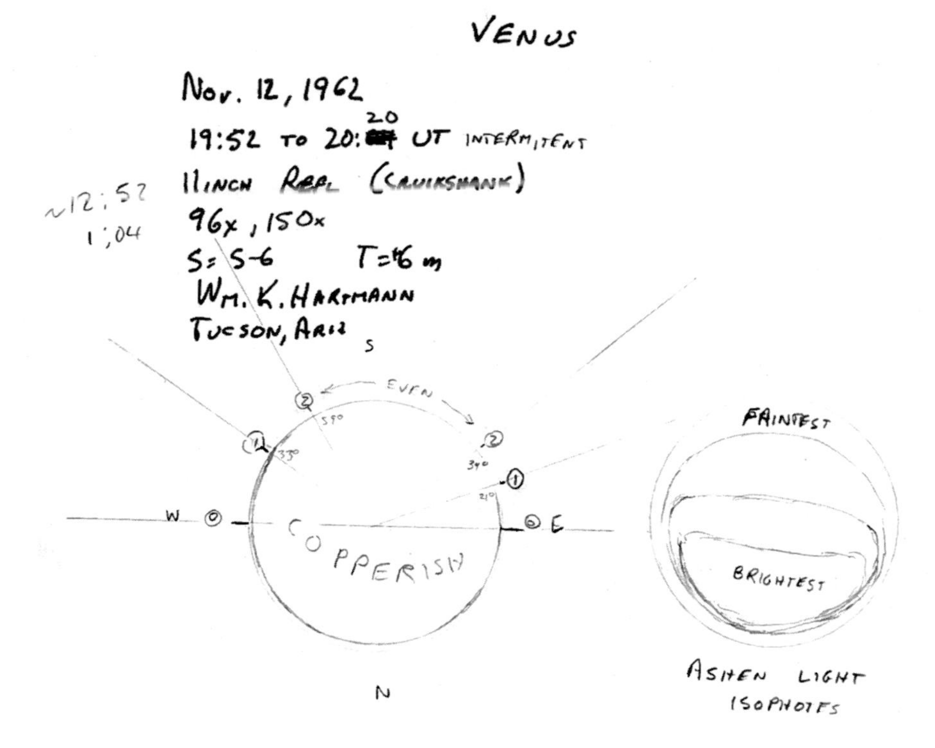

Fig. 6.5 An excerpt from Bill Hartmann's observing notebook on November 12, 1962, illustrating his observation of a discoloration on the night side of Venus relative to the surrounding blue daytime sky. Note the characterization of the effect as "copperish" in color

recollection, only to find they independently drew the same thing. Hartmann's sketch of Venus as seen on that November afternoon (Fig. 6.6) attempts to render the subtle shading he saw, coming out an odd greenish that merges with the wash of blue representing the daytime sky seen around and in front of the Venus disc. "This is my first 'real observation' of Ashen Light," Hartmann concluded in his notes.

It would also be his last such observation to date of which he is certain. Hartmann told me that although his visual Venus work dwindled considerably as he turned his graduate student attentions elsewhere, he never again saw any visual indication of anything quite like that singular observation from the fall of 1962.

Fig. 6.6 Bill Hartmann's contemporaneous colored pencil drawing of Venus as it appeared to him in daylight on November 12, 1962, the only date during the inferior conjunction passage that year on which he and Dale Cruikshank both observed a distinct 'discoloration' of the night side of Venus. Compare with both the sketch in Fig. 6.5 and the artist impression in Fig. 2.9

The Early Spacecraft Era

Along with Mars, Venus was a target of the earliest spacecraft missions beyond the Earth-Moon system. This was not only because it gave up so few secrets to Earth-bound observers, but that it was just easy to reach: the planets nearest Earth require the least amount of energy to get to, and the short trajectories of these missions were least prone to en route failure or compounded navigational errors that might well cause spacecraft to miss their targets altogether.

The Soviets made the first attempt at a Venus shot with the *Tyazhely Sputnik* mission that ended in failure in February 1961 after a problem with the launch vehicle's upper stage left it stranded in low Earth orbit. In classic Soviet disinformation style, the government announced after the failure was detected that the whole affair was the deliberate test of a "heavy satellite" launch platform and that everything had gone strictly according to plan. The Soviets' second attempt followed a few days later with the *Venera 1* spacecraft.

It made it off the ground and out of Earth orbit successfully, and along with the Soviet *Luna 2* probe discovered the phenomenon of the solar wind during its outbound flight. However, a communications failure on the way to Venus resulted in its planned mission to impact the planet's surface changed into a flyby at a distance of about 100,000 km. The first in a long line of Soviet *Venera* missions simply sailed past Venus into a solar orbit, unable to radio home information about anything it might have detected.

In an ironic twist of fate, the Americans were the first to carry out a successful Venus mission. After a failed first attempt in the series, the US *Mariner 2* probe became the first robotic explorer to carry out a successful planetary encounter as it skimmed 34,000 km above the cloud tops of Venus on December 14, 1962. Its radiometers measured cool temperatures in the clouds and a fiery 500°C at its surface, but its magnetometer sensed nothing. Given the prevailing view among scientists at the time—that Venus was probably some analog of the almost identically-sized Earth—the fact that *Mariner 2* didn't detect a magnetic field in the vicinity of the planet with anything like a terrestrial strength turns out to be most relevant in understanding what the Ashen Light probably *isn't*.

Joel Levine, later a research professor in the Department of Applied Science at the College of William & Mary, was a graduate student at New York University in 1966 when he analyzed 129 accounts of Ashen Light appearances recorded by visual observers between 1954 and 1962. Levine found that reports were most likely when the geomagnetic activity index was high around the time of inferior conjunction. His analysis was helped considerably by the large number of reported sightings in the 1950s, corresponding with one of the most magnetically active periods on the Sun in the twentieth century. Levine concluded that the Ashen Light must be an auroral phenomenon on the night side of Venus. That his study period included an unusually strong solar maximum and heightened vigilance among planetary observers during the International Geophysical Year seems to strengthen support for Levine's conclusion.

However, *individual* historical observations of the Ashen Light don't seem to be particularly well-matched with specific solar magnetic events. While there is a broad peak in the distribution of occurrences roughly centered on historic solar maxima (Fig. 2.8), Ashen Light sightings have been recorded at essentially every phase of the solar cycle. The distribution of sightings is somewhat asymmetric with a peak 1 to 2 years after solar maximum, but the correlation is weak and it is not safe to conclude that Ashen Light sightings are more likely near solar maximum than minimum.

In addition, specific attempts searching for evidence of the Ashen Light during times that might favorably correlate with elevated levels of solar activity have generally turned up empty-handed. For instance, in his report to the British Astronomical Association in December 1968, BAA Mercury and Venus Section Director J. Hedley Robinson noted that during the western elongation of 1967, the Ashen Light was not once reported by Section observers. "This is," Robinson wrote, "despite the deliberate search during high solar activity at this time, especially around 21 December 1967 and 27 January 1968."[30]

However, enough of a connection seems to exist that it's not unreasonable to believe that the Ashen Light is a real enough phenomenon. "Whatever its nature," Bill Napier wrote in 1971, "the correlation [between solar activity and reports of the Ashen Light] militates strongly against the proposition that the Ashen Light is illusory; it is difficult to believe that an observer is more prone to optical illusion on a magnetically disturbed night." But it seems safe to conclude that the Ashen Light isn't caused by a process substantially similar to terrestrial aurorae, since the planet lacks its own permanent magnetic field of any appreciable strength.

Still, early telescopic observers sometimes saw strange things on Venus that reminded them qualitatively of aurorae. In 1874, Adelbert Safarik recounted observations by Johann Heinrich von Mädler on April 7, 1833, during which Mädler "saw Venus surrounded by long bright immovable rays." Safarik also anticipated later spectroscopic searches for evidence of aurorae, writing that "Professor [Johann Karl Friedrich] Zöllner, of Leipzic, strongly advocates this idea, and trusts that the spectroscope will reveal bright lines in the grey light of the unillumined hemisphere of Venus." As we found out earlier, evidence for such light emission exists, but it turns out to be considerably fainter than what appears necessary for visual telescopic observers to detect form Earth. Later searches failed to validate the observations of Nikolai Kozyrev and Gordon Newkirk, suggesting that the spectral emission lines they recorded were improperly identified or resulted from some kind of spurious source.

While Venus may sometimes experience a weak, induced magnetic field due to the interaction of its ionosphere with the solar wind, it seems unlikely to result in aurorae of sufficient intensity to produce a glow observable from Earth; furthermore, it hasn't been seen by any of the spacecraft that have visited the planet. However, as we'll see, there may be another way to get the job done that results from a similar mechanism working in a slightly different way.

---

[30] 1968, *Journal of the British Astronomical Association*, 79(1), 52.

7

New Ideas for an Old Problem: Observations and Science 1980–2020

The 1970s were an extraordinarily productive time for Venus research, with space missions making the first direct measurements in its atmosphere and on its surface. The Soviet Union was spectacularly successful, landing a series of spacecraft that sent home the first up-close views of our neighboring world. *Venera 7* became the first spacecraft to successfully soft-land on another planet and transmit a signal back to Earth; it made in situ environmental measurements of a broiling, 475°C surface with a pressure ninety times that at sea level on Earth. The follow-up *Venera 8* mission proved that the lower atmosphere of Venus was relatively clear and that its surface rocks were compositionally similar to the basaltic rocks of our own planet. In October 1975, *Venera 9* radioed back to Earth the first photographs taken at the surface of Venus, revealing an expanse of plate-like rocks showing no evidence of erosion. Later in the decade, the Americans flew both spacecraft in the *Pioneer Venus* series, one of which continued relaying data back to Earth for almost 15 years. These missions yielded new insights into the composition and structure of the Venus atmosphere, as well as providing the first global map of the planet's surface, gleaned from radar measurements. At the same time, reported sightings of the Ashen Light fell to 50-year lows, and interest in the subject among professionals, as gauged by mentions in the scholarly literature, all but dried up. The flood of spacecraft data begat a renaissance of Venus studies, this time rooted in the fully fledged modern pursuit of planetary science. Venus had become a *place* not entirely unlike our own planet, and scientists were occupied with understanding its physical characteristics in light of emerging theories of planetary geology and atmospheres. The Ashen Light seemed in comparison

J. C. Barentine, *Mystery of the Ashen Light of Venus*, Astronomers' Universe,
https://doi.org/10.1007/978-3-030-72715-4_7

like a quaint curiosity of a bygone era, when planetary astronomers knew nothing more about planets than their telescope eyepieces and photographic emulsions could tell them.

The most sanguine proponents of Ashen Light as a real phenomenon worthy of the planetary science treatment pointed to spacecraft results as their latest, greatest, and best line of evidence. Writing in 1982, Garry Hunt and Patrick Moore argued that this new, firsthand information pointed directly at the Venus atmosphere as the origin of the Ashen Light. "We have now been able to show," they wrote,

> that there are violent electrical phenomena in the Cytherean atmosphere, with almost continuous lightning and a marked atmospheric glow. Whether or not this is responsible for the Ashen Light remains to be seen. All we can really say is that if the Light as reported by Earth-based observers is real, and is not due to mere contrast, then it is almost certainly electrical in nature.

Toward the end of the decade, the latest Ashen Light observing campaign took its cues thusly, hoping to link its appearance to conditions in the Venus atmosphere measured by our robotic emissaries.

The Phillips and Russell Campaign (1988)

The latest coordinated effort to engage astronomers in a systematic Ashen Light observing campaign came five decades after Daniel Barbier's decision to consign the phenomenon to purely legendary status (Chap. 6). Christopher Russell, a professor in the Department of Earth and Space Sciences and the Institute of Geophysics and Planetary Physics at the University of California at Los Angeles, took up the problem in the mid-1980s with his graduate student at the time, John Phillips. The latter would yet have had a highly successful career on the basis of his exploits as a naval aviator and veteran of three missions to the International Space Station, but to his already extensive résumé he added a doctorate in geophysics and space physics under Russell's supervision. At UCLA, Phillips studied data from the *Pioneer Venus Orbiter* (PVO) spacecraft and wrote a dissertation on the effects of the interplanetary magnetic field on the interaction of the solar wind with the Venus atmosphere. Between his graduate studies and visits to the ISS, Phillips was principal investigator for the Solar Wind Plasma Experiment aboard the NASA *Ulysses* spacecraft that flew multiple orbits over the poles of the Sun during 1994–2008.

During his dissertation work, Phillips began thinking about the Ashen Light. He was particularly inspired by an incident in May 1980 that suggested a correspondence between spacecraft data and Ashen Light reports from amateur astronomers. Patrick Moore reported seeing it on the evening of May 27, some 3 weeks before inferior conjunction; about 18 hours later, PVO detected the passage of a shock wave across Venus caused by the impact of a solar magnetic disturbance. Phillips and coworkers later identified similar instances in 1979 and 1982 when the impacts of solar charged particles on the ionosphere of Venus produced night-side glows in ultraviolet lines of atomic oxygen seen by PVO.[1] This suggested to Phillips a mechanism that could plausibly explain Moore's observation particularly and Ashen Light reports generally: "ions or electrons, perhaps from solar flare activity, bombard the uppermost regions of the atmosphere and create aurora-like effects, which could be related to the ashen light."[2] Phillips and Russell called on amateurs to observe the planet regularly during the 1988 Venus apparition, planning to match up observations with PVO measurements of solar and planetary conditions near the planet.

PVO included two instruments that might have detected the Ashen Light directly. Its ultraviolet spectrometer gave some indications of airglow during its primary mission, but Phillips and Russell noted that the light could not explain visual observations from Earthbound astronomers. The second instrument was a photopolarimeter, a device that measures both the intensity of light and its polarization state. One of the four channels of the PVO photopolarimeter operated at visible wavelengths and so might be used to look for optical signals around the same time that astronomers on Earth reported the Ashen Light. However, in its previous searches, the device recorded nothing. "PVO investigators …believe the instrument might have scanned across the planet too quickly for it to detect the faint glow," they wrote, "or perhaps the ashen light involves wavelengths other than those the photopolarimeter monitors. All this underscores the need for telescopic observations."

They called on participants in the campaign to report both positive and negative results, as well as "other phenomena similar in appearance but not related to the ashen light (though still of interest to us)," including the "negative" form of the Ashen Light and crescent cusp extensions. They hoped for an abundance of observations from amateurs distributed widely across longitude, for "large numbers of astronomers will also enhance the credibility of positive sightings." 1988 was an opportune time for the program: not only

[1]Phillips, J., Stewart, A., & Luhmann, J. 1986. The Venus ultraviolet aurora: Observations at 130.4 nm. *Geophysical Research Letters*, 13(10), 1047–1050.

[2]1988, The Ashen Light of Venus, *Sky and Telescope*, 75, 250.

was what turned out to be a strong solar maximum approaching, but PVO was aging and wouldn't last forever.

The campaign was successful in that it generated hundreds of reports made through telescopes ranging in aperture from two to 36 inches. "Observers viewed Venus in integrated light as well as with a wide variety of filters," Phillips and Russell wrote in 1992, summarizing the program and its outcomes. "Some observers provided detailed descriptions of seeing and transparency, and observed for several hours at a time with sequences of eyepieces, filters, and occulting devices."

British amateur astronomer Gerald North recounted a typical observer experience during the campaign:

> I took part in the programme myself and had one definite sighting of the ashen light and several suspect ones. During my positive sighting, the dim glow of the night-time hemisphere reminded me of Earthshine on the Moon, though it was less easy to see. At the time I was using three telescopes at the Royal Greenwich Observatory: a 7-inch (178 mm) f/24 refractor, a 30-inch (0.76 m) coudé reflector and a 36-inch (0.91 m) Cassegrain reflector. The ashen light was visible through all three telescopes. I saw the glow as grey in colour and another observer present saw it as blue-grey.[3]

Unfortunately for their project, the Sun just didn't cooperate; the space environment around Venus observed by PVO during the campaign was described as "unusually quiet." In the few instances where they saw evidence for solar wind shocks impinging on the Venus atmosphere, there was simply no strong correlation with Ashen Light reports. "Thus," they wrote, "we must conclude that there is no compelling evidence for a solar origin for the ashen light . . . as there appears to be little correlation between ashen light observations and observable solar particle or solar wind features at Venus."

Although their careful analysis of the reports led them to confess their inability "to unambiguously demonstrate that the emissions are real," in the end Phillips and Russell found the reports to be sufficiently believable to keep some faith in the possibilities associated with what remained for them an unsolved problem. "We encourage astronomers to continue efforts to solve this centuries-old riddle," they signed off hopefully in 1992, anticipating that the advent of low-cost digital detectors might put into the hands of amateur astronomers devices with the sensitivity and high dynamic range that could definitively answer the otherwise still-unanswered question.

---

[3] 1997, *Advanced Amateur Astronomy* (2nd ed.), Cambridge, UK: Cambridge University Press, 189.

A quarter century later, I asked Professor Russell whether he thought at the time that the study offered something more substantive. With the benefit of years' worth of hindsight, he was more confident in his answer. "For me it did settle the question," he said in response, noting that many of the reports received during the 1988 campaign were made under relatively poor seeing conditions. He and Phillips just did not find the "smoking gun" evidence that would have resolved the controversy, and certainly experienced nothing like the 1980 episode in which it seemed like Patrick Moore's faith in the reality of his Ashen Light observations over the decades was validated by the PVO data. The connection should have become more clear during the coordinated program in 1988, but the evidence simply failed to materialize. In the end, it's difficult to argue against Russell's logic for the persistence of belief in the Ashen Light: "With any controversy there are adherents who tenaciously hold on to their position forever."

Lightning in the Venusian Dark

The ideas about the Ashen Light that have persisted the longest are those with clear terrestrial analogues: if the Ashen Light is a real physical phenomenon, and given that simpler explanations tend to be correct, then the right answer may well be found right here on Earth. From here to the end of the chapter, we'll look at some possible explanations that grow increasingly likely based on the available evidence.

One way to light up the night side of a planet like Venus is through cloud-to-cloud lightning in an atmosphere thick enough to generate electrical storms, yet thin enough to permit visible light to escape easily. With a sufficiently high rate of electrical discharges in the clouds, provided that the discharges were fairly widespread throughout the atmosphere, an amount of light may be generated that would yield a weak but more or less uniform glow. The irregular frequency of appearance of the Ashen Light might therefore be indicative of seasonal variations in the intensity of storms. The lightning hypothesis also makes testable predictions concerning both the direct detectability of lighting and indirect effects on the chemistry of the Venusian atmosphere.

In fact, early investigations of the spectrum of the *daytime* side of Venus revealed traces of certain molecular compounds like nitric oxide (NO) and nitrogen monohydride (NH) whose formation was thought to involve electrical discharges. However, given that there are other ways to make these molecules, scientists looked to directly sense the electromagnetic signals of lighting. Some early spacecraft missions to Venus detected flashes of light on the planet's night side, beginning with *Venera 9* in 1975. When the Jupiter-bound *Galileo* spacecraft flew past Venus in 1990 to pick up speed in a gravity-assisted maneuver, mission controllers turned its science instruments toward the planet and recorded some faint optical emissions consistent with lightning.

Yet other missions such as the *Pioneer Venus Orbiter* (1978) and the Soviet *Vega* balloon experiment (1986) saw nothing. In 1990, Christopher Russell and Fred Scarf explained these non-detections with an analogy to lightning on the Earth driven by convective thunderstorm development, which tends to peak in a given location during the later afternoon hours. *Venera 9*, they argued, observed Venus in a direction corresponding to the expected activity peak, while the other spacecraft observed parts of the planet where it was early morning when the lighting incidence rate should be low. And anyway, they noted, neither *PVO* nor *Vega* made use of instruments especially well-suited to detecting optical signals from lightning.

While searches for emissions of visible light from Venusian lightning yielded ambiguous results, efforts targeting other parts of the electromagnetic spectrum had better luck. In addition to visible light, lightning emits radio waves; anyone who has listened to AM radio during a thunderstorm and heard the faint crackles of static correlated with distant flashes of lightning has experienced this firsthand. These so-called "spherics" can often be heard hundreds of kilometers away from the lightning that produces them. *Cassini*, another American spacecraft bound for the outer planets, flew past Venus twice, in 1998 and 1999. Mission scientists looked for evidence of spherics in data from *Cassini*'s radio and plasma wave science instrument, initially convinced they had turned up nothing even though the same instrument detected as many as 70 individual lightning discharges per second when *Cassini* flew by Earth on August 18, 1999.

Researchers were puzzled and began to think that maybe Venusian lightning just doesn't behave the same way lightning does on Earth. "Since the atmosphere of Venus is very different from that of Earth," said University of Iowa physicist Donald Gurnett, who led the *Cassini* study, "it is perhaps not surprising that electrical activity on Venus might be very different from lightning in the Earth's atmosphere."[4] In 2007, early results from the European Space Agency *Venus Express* spacecraft found evidence for a certain kind of electromagnetic wave known as a "whistler" propagating from the lower

[4]Quoted in "Cassini Scientists See No Sign Of Lightning On Venus", NASA press release #2001–013, January 19, 2001.

atmosphere to the ionosphere around 100 times per second. These bursts of radiation, lasting from a quarter to a half-second, are consistent with a lightning origin.

The preponderance of the evidence suggests that lightning is indeed a regular feature of the Venusian atmosphere. What keeps it from being a leading contender for the source of the Ashen Light is that the implied frequency of the flashes doesn't yield sufficient optical radiation to make the planet's night side glow brightly enough to be visible from Earth. In 1992, Dale Cruikshank estimated that the rate of lightning strikes sufficient to produce a visually detectable glow is on order 1,000 per second, about ten times the instantaneous rate of cloud-to-ground strikes on the nighttime hemisphere of Earth. A lightning strike rate on the night side of Venus high enough to explain the Ashen Light would have been easily detected by orbiting spacecraft and atmospheric balloons. Furthermore, the frequency of lightning implied by the *Venus Express* observations does not account for the episodic nature of historical Ashen Light observations. Unless there is some characteristic of Venusian lightning that makes it very different from the familiar variety here on Earth, we can probably safely dismiss it as a possible source for the Ashen Light.

Active Volcanism

As a postscript to the lightning theory, Cruikshank put forward the idea that intense lightning could result from active volcanism on Venus, since

> lightning on Earth often accompanies violent volcanic eruptions. In those events (e.g. Surtsey on Iceland in the 1960s), rapidly rising columns of dust, gas, and ash produce the electrical charge separation that eventually leads to lightning flashes. One could, therefore, speculate that large-scale volcanic eruptions on Venus's night hemisphere could produce lightning discharges that are visible through the clouds and visible from Earth.

While there is no irrefutable evidence that active volcanism takes place now on Venus, there are some tantalizing hints in spacecraft data that the planet might not be geologically dead. If it were the case that lightning on Venus indicated a geologically living planet with active volcanoes, it might support the idea of Ashen Light as thermal emission from the surface. Although the surface only very faintly emits light at visible wavelengths, volcanically active regions on Venus would be much warmer than their surroundings. This would have the effect of shifting some of their emissions into the range of wavelengths readily observable by the human eye. A promising way of locally heating up the surface is through active volcanoes, whose vents might spew lava at temperatures upward of 800 °C.

The idea of active volcanoes on Venus isn't new, having first been raised scientifically in print a half century ago; in fact, the planet's high surface temperature was once taken as suggestive of both volcanic and tectonic activity. Writing in the late 1960s, Gerald Davidson and Albert Anderson noted that spacecraft confirmation of a hot Venus "raised much speculation about atmospheric processes," referring to the hypothesis of a runaway greenhouse effect, "but little effort has yet been expended toward reconciliation of the observations with the theory of planetary interiors."[5] Instead of finding an explanation in the clouds above the surface of Venus, they argued, the answer might be due to the welling up of heat from below:

> A high surface temperature would imply that the surface heat flow on Venus is considerably greater than on Earth, so temperature gradients would be correspondingly increased. A surface temperature of 500°K and a thermal gradient (say) twice that of Earth would result in temperatures of 1300°K at depths less than 25 km, which would be adequate to melt silicate rocks. Therefore the crust is probably quite thin; it may even float on a layer of molten rock.

Furthermore, if volcanoes on Venus were anything like those on Earth, they were probably belching certain telltale gases into the atmosphere, and maybe even *making* the atmosphere itself. A single eruptive event from a terrestrial volcano can emit millions of tons of carbon dioxide into the Earth's atmosphere, and sure enough, CO_2 is the main constituent of the atmosphere of Venus. Terrestrial volcanic eruptions also inject molecules and dust into the atmosphere, reaching heights of 15 to 20 km above the surface. Atmospheric chemistry involving sulfur dioxide (SO_2) molecules of volcanic origin drives formation of sulfate aerosols in a process similar to that resulting in the formation of particulate hazes over cities.

Davidson and Anderson speculated that a complementary mechanism might account for "much of the atmospheric obscuration of Venus." Indeed, analysis of ultraviolet spectra obtained by the *Pioneer Venus Orbiter* spacecraft in the late 1970s and early 1980s showed a steep decline of both the abundance of SO_2 and submicron-sized sulfate aerosol particles in the upper atmosphere

[5]1967, *Science*, 156(3783), 1729–1730.

relative to upper limits measured in the 1960s.[6] This may indicate a quieting-down of an extensive and ongoing eruptive episode somewhere on the planet.

More recently, direct detection of apparent hotspots on the surface of Venus from orbit offers perhaps the strongest evidence for ongoing volcanism. In 2015, a team of researchers led by Eugene Shalygin of the Max Planck Institute for Solar System Research analyzed bright spots seen in images obtained with the Venus Monitoring Camera onboard the *Venus Express* spacecraft. The locations glowed brightly in light with a wavelength of 1.1 microns, implying a temperature considerably higher than their surroundings. The spots ranged up to 200 square kilometers in size, and their temperatures were around 300° higher than the Venus global mean. From multiple observations spaced apart in time, the authors concluded that the spots "correspond to volcanic eruptions and related changes in surface temperature due to eruption of lavas."[7]

Five years later, geophysicist Anna Gülcher and her colleagues re-examined historical data from spacecraft missions, including synthetic aperture radar altimetry from the American *Magellan* mission of the early 1990s, and compared observations to thermomechanical models of the interior of Venus. In particular, they focused on peculiar circular structures called coronae, thought to form when plumes of hot magma rising through the mantle of Venus push into the planet's crust, causing it to deform upward into a dome shape; as the magma cools near the surface, the dome's center collapses. The resulting structures take their name from the Latin word for "crown" on the basis of their three-dimensional appearance with a raised rim and depressed center. Gülcher and coworkers concluded that at least 37 coronae on Venus have been "very recently active."[8] This strongly suggests that the Venus interior is still convecting, if not driving volcanism to this day.

Active volcanoes whose eruptive fields reach temperatures upward of 1000 °C could emit enough visible light to be seen from Earth if viewed through unusually thin cloud layers that create temporary "windows" through which the radiation can escape. In this scenario, unpredictable volcanic activity and intermittent transparency of the Venus atmosphere to optical radiation join forces with the irregular sampling of Venus in time by telescopic observers

[6]Esposito, L. 1984. *Science*, 223(4640), 1072–1074. Esposito concluded from the available data that "the explanation of this phenomenon was a single injection of SO_2 into the Venus middle atmosphere prior to the Pioneer Venus encounter in 1978. … A natural physical mechanism for this injection is major volcanic activity on the surface."

[7]2015, *Geophysical Research Letters*, 42(12), 4762–4769.

[8]Gülcher, A., Gerya, T., Montési, L., Munch, J. 2020. Corona structures driven by plume–lithosphere interactions and evidence for ongoing plume activity on Venus. *Nature Geoscience*, 13(8), 547–554.

to yield an explanation that accounts for several important aspects of the Ashen Light phenomenon. But to the extent that the volcanoes might well also emit materials that simultaneously *increase* the opacity of the atmosphere, the volcano theory becomes less tenable. That brings us back to perhaps the best place to look for a physical cause of the Ashen Light: high above the planet in its upper atmosphere.

Airglow Revived: New Observations of the "Green Line"

In 1975 the Soviet *Venera 9* and *Venera 10* spacecraft detected faint visible light on the night side of Venus, attributed to emission from excited oxygen molecules, that ground-based spectroscopists evidently overlooked. Maybe, on occasion, could this light become unusually bright and telescopic observers see it as the Ashen Light?

Several years later, the Soviet astronomer Vladimir Krasnopolsky pointed out that in order to be visible from Earth, the strength of the emission would have to be as much as one hundred times brighter than that measured by the *Venera* spacecraft. When I reached him by email during the research for this book, Krasnopolsky was quick to point out that "the term 'Ashen Light' is not currently used in the Venus science." The proper term in the research community these days is "Venus night airglow," he said, and it is already well-established in the forms of molecular oxygen in the visible spectrum and at 1.27 microns in the infrared; nitric oxide emission in the ultraviolet; and glow from hydroxyl ions near 3 microns. The pride in his words about his work on the Soviet *Venera* spacecraft was evident: "Along with the first images of the Venus surface, the discovery of the Venus airglow was an impressive result of Veneras 9 and 10." But he admitted that further ground-based searches for low-level optical Venus atmospheric glows through the early 1990s had turned up nothing. Like other theories purporting to explain the Ashen Light throughout history, it seemed like atmospheric emission was yet another dead end.

But new evidence shortly after the turn of the twenty-first century brought these ideas back to life. Underlying the tantalizing detections (and failures to detect) visible light from the night side of Venus was a belief that there might well be a local analog to the phenomenon of optical "nightglow" in the Earth's atmosphere. The concept goes like this: atoms and molecules high in our atmosphere are exposed to intense ultraviolet light while they are on the daytime side of the planet. Absorbing solar ultraviolet photons changes their internal energy configurations in different ways that cause them to emit visible light. An atom might get enough of an energy kick from absorbing an ultraviolet photon that it enters an excited state, but the rules of quantum physics prevent it from staying that way for very long. It can get rid of that extra energy if it collides with another atom or a molecule, but if it manages to avoid that fate it will simply convert the energy to light and emit a photon with a characteristic wavelength.

Incoming solar ultraviolet photons with even higher energies can peel off an atom's outer electrons entirely, putting it into an ionized state that persists until the atom encounters free electrons and "recombines" into an electrically neutral condition. Similarly, solar ultraviolet light can excite molecules or dissociate them entirely by splitting them apart into their constituent atoms. Those constituents may recombine with atoms of the same kind or different kinds, forming new kinds of molecules and contributing to the diversity of chemistry in the upper atmosphere. Given the flux of solar photons on the daytime side of the planet, atoms and molecules are constantly being excited, ionized, and dissociated, and subsequently recombining and de-exciting. On the sunward-facing side of Earth, this situation more or less reaches an equilibrium and causes a steady emission of light into the daytime sky, but we don't see it because the scattering of sunlight from nitrogen molecules results in a background level orders of magnitude brighter than airglow.

Some of those excited atoms and molecules stay ionized or dissociated long enough that high-altitude winds and the rotation of the Earth carry them around to the night side of the planet where they can find each other and recombine. Since energy was "borrowed" from the solar ultraviolet photons to change the energy states of the atoms and molecules, and because energy must be conserved, some of that energy is returned in the form of visible light when recombination takes place. But it's not exactly the same amount of energy, because, for example, during dissociation of molecules, some quantity goes into changing the internal energy of the constituent atoms. Since the energy returned upon recombination is less than that of the original ultraviolet photon, a new photon can be emitted that has a wavelength in the part of the spectrum to which the human eye is sensitive. If a sufficient number of these events are happening in a particular direction on the night sky, the resulting airglow can be bright enough to become visible to the unaided eye (Fig. 7.1). Other instances return an ultraviolet photon of a different energy than the one that dissociated the molecule in the first place; in either case, some detectable light is produced. If something similar is going on in the atmosphere of Venus, it might explain both the visible light and the ultraviolet emission seen by spacecraft.

Fig. 7.1 Green and red airglow over the village of Kinnulanlahti in Northern Savonia, Finland, on September 3, 2013. Yellow light nearer the horizon is caused by light pollution from the village. Image by Flickr user Janne, licensed under CC BY-SA 2.0

Similarly, individual atoms in the atmosphere can pick up some excess internal energy, either through irradiation with solar ultraviolet light or by collisions with other atoms or molecules. The process of producing an aurora is substantially the same, except that the colliding body is in that case a solar charged particle rather than an atom or molecule. It's not surprising, then, that the strongest emission of visible light in the nighttime atmosphere comes from the materials that make up most of our air: nitrogen and oxygen. The latter is the source of two of the strongest emissions: green light at 557.7 nanometers (nm) wavelength and red light at 630.0 nm. If a sufficient amount of free oxygen exists in the atmosphere of Venus, it stands to reason that we might see the "green line" and the "red line" features with a sufficiently sensitive telescope and detector.

In November 1999, Tom Slanger, then a senior staff scientist at the SRI International Center for Geospace Studies, trained the 10-meter W.M. Keck I telescope on Venus to look for the green and red oxygen lines in the most sensitive search to date. Not only did Slanger detect the green line on the night side of the planet, but he found its strength to be comparable to the strength of the same feature in our own planet's atmosphere.

Curiously, the team *didn't* see the red line, which offered an important clue about where the light was coming from. While they detected the green line, they didn't know at the time the altitude at which the light was emitted. Energetically speaking, it's a lot easier to make red-line light than green-line light, so the absence of red-line emission indicated that the green line must be forming at an altitude below 150 km; otherwise, physics predicts about five times more red than green. Below 150 km, the Venus atmosphere becomes sufficiently dense that the red line is "quenched" by collisions between excited oxygen atoms and other gaseous constituents of the atmosphere. The green line, on the other hand, is relatively unaffected by this process at similar altitudes.

Slanger and coworkers therefore hypothesized that the large discrepancy between red-line and green-line emission was attributable to a scenario in which the green-line light comes from fairly deep in the atmosphere—not in the ionosphere of Venus, a very high-altitude atmospheric layer. If more red light had been seen, it would indicate an origin consistent with aurorae; instead, it seemed like Slanger's team had something in hand more akin to terrestrial airglow (Fig. 7.2).

Slanger's result starts to resemble a familiar pattern: oxygen molecules are split up by solar ultraviolet radiation on the day side of Venus and are quickly carried by strong winds to its night side. There, they encounter each other and recombine to make oxygen molecules with internally excited energy states. If those molecules happen to collide with electrically neutral oxygen atoms in the lowest possible energy state, the atoms can then be bumped up to excited states. After a short while, the excited oxygen atoms "drop" into the ground state, emitting light in the green line. At higher altitudes, free oxygen atoms can enter excited states after encounters with ultraviolet photons so that they have a net positive charge. Some of those atoms will find free electrons, combine with them, and give back their energy in the form of green-line light.

But then what should we make of reports through history that give the Ashen Light an uneven, blotchy appearance, suggesting non-uniformity of distribution around the night side of Venus? If the source of the Ashen Light were like airglow and the atmosphere of Venus behaved at all like our own, the answer might have something to do with so-called "bright nights" on Earth. Since antiquity, observers have commented on nights when the sky appeared unusually bright, even in the absence of the Moon, aurorae, or otherwise familiar sources of natural light. On these nights, light from the sky is described as bright enough to read by. In Chap. 4, we encountered such a description by the German astronomer Karl Ludwig Harding, who noted that on these

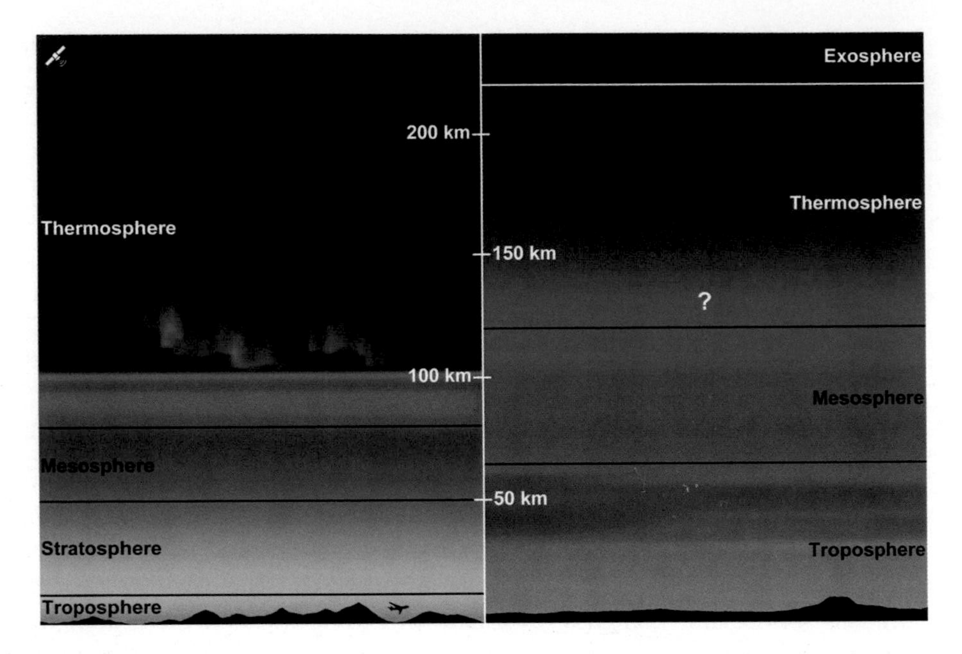

Fig. 7.2 Comparison of the Earth (left) and Venus (right) atmospheres at the same altitude scale. In the terrestrial atmosphere, sodium airglow forms at the bottom of the mesosphere, with a thin layer of oxygen "green-line" airglow above it. At altitudes above 150 km, the oxygen "red-line" airglow forms; between them is where the aurorae occur. On Venus, oxygen green-line emission forms below 150 km, but its exact altitude and extent are still unknown. The planet's dense clouds and haze layers form lower in its troposphere

nights "distant objects can be recognized to me quite distinctly." Perhaps the most frequently quoted reference in literature comes from *Naturalis Historia* ("Natural History") by the first century CE Roman author Pliny the Elder: "What most people call 'suns at night,' light from the night sky is seen as during the consulship of C. Cæcilius and Cn. Papirius and at many other times, and gives the appearance of day in the night."[9]

It was in fact a bright night in 1929 that led to the identification of airglow as a principal contributor to the brightness of the night sky by Robert John Strutt, 4th Baron Rayleigh.[10] Rayleigh's spectroscopic observations showed that a factor of four increase in the brightness of the night sky seen on the night

[9] *Naturalis Historia*, 2.33. Gaius Caecilius Metellus Caprarius and Gnaeus Papirius Carbo were Roman consuls in the year 113 BCE.

[10] Posthumously nicknamed the "Airglow Rayleigh" to distinguish him from his father, John William Strutt, 3rd Baron Rayleigh (1842–1919), known as the "Scattering Rayleigh" for his work on the scattering of light by small particles.

of November 8 was not due to the aurora.[11] Further investigation revealed that these airglow enhancements were observed regionally but not globally; in other words, they had some sort of spatial structure distributed in longitude.

More than eight decades would pass before a plausible explanation emerged, supported by evidence: density waves passing through the ionized upper atmosphere could converge over a particular region, raising the brightness of the airglow locally. In 2017, Gordon Shepherd and Young-Min Cho of the York University Centre for Research in Earth and Space Science published observations from the Wind Imaging Interferometer (WINDII) instrument aboard the *Upper Atmosphere Research Satellite* showing how the terrestrial oxygen "green-line" airglow varied with longitude and corresponded to the meeting of atmospheric pressure waves.[12] These waves are characterized by their zonal wavenumbers, dimensionless figures equal to the number of whole wavelengths that fit within a circle around the globe at a given latitude. "It is concluded," the authors wrote, "that the historical bright nights are consistent with their identification as enhanced airglow and that these enhanced airglow events arise from the occasional superposition of the peaks of zonal waves 1–4, giving rise to localized enhancements of emission rate by a factor of about 10 over minimum airglow levels." For an assumed human visual detection threshold of 200 Rayleighs, "the bright nights observed by WINDII are definitely detectable with the human eye and … they very likely produced the visual effects seen by historical observers."

Might something substantially similar happen in the Venus atmosphere? Planetary-scale atmospheric waves are known to exist on Venus; in fact, these processes are thought responsible for the distinctive, global "Y"-shaped cloud pattern seen in ultraviolet photos (e.g., the left image in Fig. 2.1).[13] The amplitudes of these waves might even be larger on a planet with a strongly "superrotating" atmosphere like Venus, leading to spatially large structures like the "Y" clouds.[14] Shepherd and Cho found that terrestrial atmospheric conditions resulting in unusually bright oxygen airglow lasted in some cases for several consecutive days; this, too, qualitatively matches historical Ashen Light reports in which one or more observers claimed to have seen the phenomenon on several successive nights. They also showed circumstances in which airglow light emission rates reached as high as 400 Rayleighs—twice the presumed

---

[11] 1931, *Proceedings of the Royal Society of London. Series A, Containing Papers of a Mathematical and Physical Character*, 131(817), 376–381.

[12] *Geophysical Research Letters*, 44(13), 7036–7043.

[13] Covey, C., & Schubert, G. 1982. *Journal of the Atmospheric Sciences*, 39(11), 2397–2413.

[14] Peralta, J., et al. 2015. *Geophysical Research Letters*, 42(3), 705–711.

visual detection threshold on Earth. If the Venus airglow were really this bright, it would be readily observable from Earth by dark-adapted observers using small telescopes.

Chasing down this hypothesis brought me to my friend Candace Gray, a member of the scientific staff at Apache Point Observatory (APO) near Cloudcroft, New Mexico. Gray started her graduate school career in 2006 with me in the entering class of first-year students at the University of Texas at Austin. Originally from El Paso, Texas, that's where I first met her, in the departures lounge of the El Paso airport, waiting on a flight to Austin. We were both at that point prospective students invited to visit the UT-Austin campus as part of deciding whether or not to enroll; for me, El Paso turned out to be the nearest airport from which I could catch such a flight. She was then a bright-eyed, enthusiastic, recent physics graduate from the University of Texas at El Paso, and the day I first met her she was wearing a t-shirt emblazoned with the phrase PHYSICS IS PHUN.

We sat next to each other on the flight and chatted the entire 2 hours en route to the Texas capital. Unlike a lot of professional astronomers, Gray wasn't one who started out as an amateur or "always knew" that she wanted to do astronomy. "I loved stargazing with my dad out in the desert and, when college came around, I signed up for an astronomy class since it sounded like fun," she later told me. "The first day, I was hooked."

Gray worked on the chemistry of comets for the first two years of graduate school before switching subjects—and universities—to pursue something that became closer to her heart: the atmosphere of Venus. At her new academic home, New Mexico State University, her thesis adviser offered her a range of potential thesis topics, including Venus. While interested, she acknowledged that the subject was a little out of her wheelhouse. The figure who made the most difference to her success in this new subject was Tom Slanger. "Tom, while huge in his field, was one of the nicest people I've ever come to work with," she said. "He always had time for me, answered all my questions with patience, and would kindly repeat answers when I asked the same question multiple times. I had so much fun working with him that, before I realized it, it was clear I'd chosen the Venus aeronomy project."

Aeronomy is the study of the upper reaches of an atmosphere, whether the Earth's or that of another planet. Part meteorology, part physics, and part atmospheric science, it relates to the bulk motion of materials in upper atmospheres; their physical conditions, chemical compositions, and properties; and how they react to the space environment. Aurorae and airglow fall under this heading along with a number of other phenomena. In the case of the Earth and Venus, being similar terrestrial planets with thick atmospheres, there's a lot of information gleaned from decades of intense study that can be applied to solve problems in the atmosphere of our neighbor world.

For Gray, her "a-ha" moment came in the fall of 2011 at a planetary science conference in France. The European Space Agency *Venus Express* science teams were well represented among the attendees. "I remember sitting in on one talk when the speaker, Thomas Widemann, said 'ignore these data, a solar flare went off and overloaded the instruments.'" But she started to put the pieces together while listening to the presentation; as she later related it, things started to "click" in her mind. It began with the realization that the oxygen green line is dominant among the emissions features in the spectra of both terrestrial airglow and aurorae.

Early observations of the Venus green line revealed that its intensity varied in time to an extent making it more like the Earth's aurora than its airglow. Because irradiation of the terrestrial atmosphere by solar extreme-ultraviolet light drives the airglow, it emits light persistently throughout the solar cycle. On the other hand, aurorae are highly episodic and vary in both intensity and geographic distribution, capable of intense emission on one night and none the next. That's because aurorae are driven by the dynamic plasma of the solar wind as it interacts with the Earth's magnetic environment. While Venus exhibited a diffuse global green-line emission like terrestrial airglow, its highly variable nature at least qualitatively mimicked the aurora.

But, as Gray well knew, Venus doesn't have a permanent magnetic field to drive polar aurorae. Was this some kind of diffuse, global aurora occurring deep in the ionosphere, unique to non-magnetic planets? The *Venus Express* results suggested that something big happened in the Venus ionosphere when it was walloped by the aftermath of a solar flare. And it was worth following up. She was sure someone must have investigated this already, but there was essentially nothing about it in the scholarly literature. "I think that's when I knew this was going to be my thing. And I was so excited!"

Given that solar flares and coronal mass ejections (CMEs)—huge releases of plasma and accompanying magnetic fields from the Sun—generate aurorae on Earth, and that the direction of the interplanetary magnetic field at the time of a CME impact is important as to whether strong aurorae tend to result, Gray looked back at all the Venus green-line data then published, sorting the observations according to observing date and where in its orbit Venus was at the time. Then she started comparing that information to records of solar activity through the same time periods. That's where things got interesting. "Every single green-line detection was preceded by a solar flare or a CME," she said. "How had no one made this connection before?!" Then there is the matter of what would make a convincing, open-and-shut case. "What I needed to do was observe Venus after a solar flare or CME and see if green-line emission was present. Since solar flares are much more common than CMEs, I tried chasing flares."

The following year, Gray was able to collect several nights' worth of Venus measurements after solar flares; however none of them showed green-line emission. While preparing a poster presentation for another academic conference, she looked back at some earlier data taken during what she thought were quiescent solar conditions. Through an accident of telescope pointing, she saw the oxygen green line near the planet's limb, while it was much less prominent elsewhere on the disc. That suggested a limb brightening effect (Chap. 2), which in turn implies a more or less uniformly glowing atmosphere made brighter near the edge of the planet's disc by long path lengths probed by sightlines in that direction. "Everyone thought the green line should be concentrated at Venus local midnight because that's where nightglow is concentrated. But either because of telescope bounce or maybe I had centered the spectrograph slit wrong, I had placed the slit on the edge of the planet."

It was another moment where a piece of the puzzle fell into place: the light emission must be happening more or less all over the planet's atmosphere at once, and that sounds a lot more like airglow than aurora. But why was it happening? Looking back at the solar activity in the days preceding her data collection showed no flares present, but there was a small coronal mass ejection aimed at Venus. That the impact of a CME immediately preceded the green-line emission pointed to charged particles being the source. On Earth, that's a lot more like an aurora than the mechanism that powers the airglow, but Venus doesn't have a strong intrinsic magnetic field to drive something like an aurora. What was going on? Gray needed a mechanism to get the charged particles to the night side of Venus in the absence of a permanent magnetic field.

Around the same time, a team led by Tianjiao Zhang of Xi'an Jiaotong University in China reported magnetic reconnection events in the space plasma environment around Venus that can explain why the planet shows indications of green-line activity even though Venus lacks a strong permanent magnetic field. In their model, solar coronal mass ejections are the source of these electrons. CMEs are gargantuan releases of plasma and open magnetic field lines from the Sun that sometimes follow solar flares. These field lines can find each other in the vicinity of Venus and "reconnect," pulling free electrons from the plasma along with them. The rain of electrons into the depths of the Venus atmosphere results in something very much like aurorae on Earth while not requiring the existence of a permanent planetary magnetic field. Unlike terrestrial aurorae, the Venus phenomenon isn't confined to its polar regions,

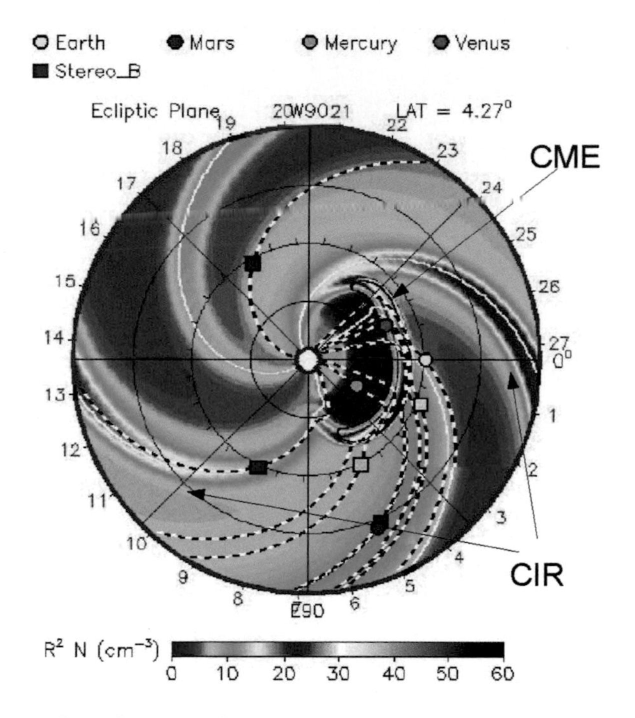

Fig. 7.3 Solar wind conditions in the inner Solar System on July 14, 2012, when a major coronal mass ejection event impacted Venus (green circle). The view shows a simulation of the solar wind density (colors) in a region of space passing through the plane of the Solar System; the vantage point is looking down from over the solar north pole. The coronal mass ejection ("CME") is the large, backward "C" shape to the right of the Sun, represented by the white dot at the center of the circle. Co-rotating interaction regions in the solar wind are the spiraling areas marked "CIR". Interplanetary magnetic field lines sensed by spacecraft are shown as black-and-white dashed lines

perhaps accounting for some amount of green-line emission in the atmosphere all over the planet.

"The fact that only the limb exposure showed emission highlighted the fact that I had been looking in the wrong place," Gray said. She hastily completed the analysis and showed the results to her thesis advisor, Professor Nancy Chanover. "When I was done, Nancy said 'sounds like you have a poster to rewrite' to which I responded 'starting on it now!!' "

The highlight of her presentation became a centerpiece of her thesis research. On July 14, 2012, a strong CME impacted Venus head-on (Fig. 7.3) in what later became known as the "Bastille Day Event." It was a serendipitous moment for Gray's work during which several essential elements suddenly converged, including an active Sun, clear weather at the observatory, and avail-

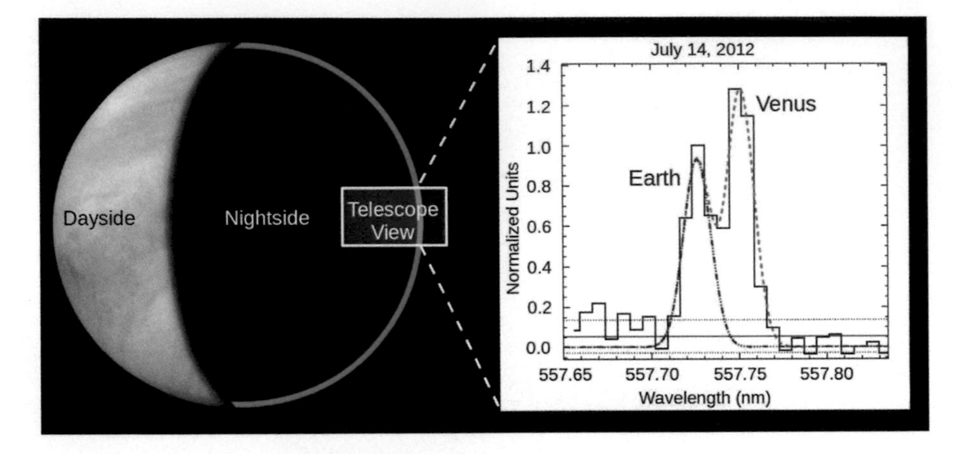

Fig. 7.4 Candace Gray's spectroscopic observations of the Bastille Day Event on July 14, 2012, are shown at right along with a cartoon representation of the Venus airglow at left. The box labeled "telescope view" indicates the position and fraction of the Venus disc sampled during the observations

ability of the telescope. She remembers being nervous during the observations, because those factors might not again line up so perfectly during the rest of her graduate career. Would she see anything in the Venus atmosphere that night? And what if she saw nothing—then what? What else could possibly explain her other data?

As soon as observations finished at dawn, Gray started a computer program to correct and calibrate the data. "I waited with bated breath for the hour it took to run its course," she said. Then, the moment of truth: "I opened the spectrum and clicked over to the spectral order containing the position of the green line. And I burst out crying." The green line in the Venus atmosphere was brighter than the same feature in our own atmosphere. "Nothing this bright had been seen before, not even by Tom. After sobbing for a bit, I collected my things and drove home to try to get some sleep." Her main result is shown in Fig. 7.4 as the spectrum of the Venus night-side limb around the expected wavelength of the oxygen green line. The Earth's green line also appears near the same wavelength, slightly shifted toward a smaller wavelength due to the relative motion between Earth and Venus at the time.

Gray finished her Ph.D. degree at New Mexico State University in 2015 with a dissertation titled "The Effect of Solar Flares, Coronal Mass Ejections, and Co-Rotating Interaction Regions on the Venusian 5577 Å Oxygen Green Line." In it, she showed that the impacts of solar charged particles on Venus correlated with night side green-line emission from the oxygen airglow, and from this she concluded that the "precipitation" of charged particles into the lower altitudes of the Venus ionosphere was primarily responsible.

Her work may help cement the explanation for why we see the oxygen green line on Venus but not the corresponding red line: "The only way I could explain this is if high energy particles are impacting directly into the deep atmosphere," she told me. "So now the story continues to find out what kinds of particles are doing this. My hypothesis is that they are high energy protons and the green line is essentially a special kind of 'proton aurora.'"

It turns out that solar wind electrons aren't sufficiently energetic to get the job done. They just don't seem to penetrate deeply enough into the Venusian ionosphere to induce oxygen atoms to give off green-line light. On the other hand, *protons* from the solar wind can. In 2019 Gray proposed that solar wind protons spiraling along interplanetary magnetic field lines draping around Venus can impact the planet's night side and enhance the electron density in the lower ionosphere around 125 km above the surface. As more electrons with the right energies become available at this altitude, more oxygen atoms are excited, and more green-line photons are emitted. And indeed, this is how the process works in our own atmosphere. But, Gray points out, what's colliding with oxygen atoms and leaving them in excited states is unclear. It could be protons impacting directly onto the oxygen atoms, a "cascade effect" of multiple particles and atoms, or something else. We don't yet know.

The circumstantial case for the Ashen Light as an airglow phenomenon seems strong. But there's one important observational fact about the Ashen Light among those recounted in the list at the start of Chap. 5 remaining unaccounted for: it is seen at all phases of the solar cycle, even at times when the Sun is magnetically "quiet." Without the benefit of a permanent magnetic field with which to wrangle the solar charged particles thought to be responsible for powering a Venus-wide airglow, the argument favoring this explanation for the Ashen Light might be in danger of collapse. But there's a different route to the same result that doesn't require the Sun to be especially active. Gray's work helps explain this as well.

The Sun constantly spews a steady but usually low-level fountain of charged particles into space. Turning on its axis once every 25 days or so, our star is like a slowly rotating garden sprinkler, sending streams of particles toward the planets along gracefully curving spiral arcs (Fig. 7.3). These solar wind streams occasionally pile up and interact with each other in so-called co-rotating interaction regions, or CIRs, locally raising the density of charged particles. They sweep slowly across the planets and so represent a longer-lasting source of energetic electrons than the induced magnetic fields resulting from CME impacts. Gray's observations confirmed that CIR impacts on the Venus atmosphere excite the oxygen green-line emission, but not quite to the same degree as do the direct effects of solar coronal mass ejections.

Of course, as is often the case with cutting-edge scientific research, there's still a lot we don't know. Space weather has significant impacts on planetary atmospheres, much of which we don't understand very well. For Venus in particular, the "superrotation" of its atmosphere relative to the planet's rotational period yields complex flow patterns that suggest a coupling of the upper atmosphere to the solar wind. "This would not be unique to Venus; we would expect this on other planets as well," Gray said. "However, the fact that Venus has such a thick atmosphere, no magnetic field to protect it from the solar wind, and is so close to the Sun combine in perfect conditions to study this interaction."

To the Limits of Human Visual Perception

Even if a plausible mechanism for generating light in the upper atmosphere of Venus were shown conclusively to become bright enough to be seen from Earth, how sure can we be that human observers on our planet would actually *see* it? To answer that question requires traveling back in time nearly 75 years to nearly the dawn of modern vision science.

Harold Richard Blackwell was born on January 16, 1921, in Harrisburg, Pennsylvania. He earned a Ph.D. in psychology from the University of Michigan in 1947, and during his graduate training in 1943, he became a resident psychologist at the Polaroid Corporation. During World War II, he headed the vision research division of the Louis Comfort Tiffany Foundation on a project designed for military applications. After the war, he continued his service to the US government, serving as executive secretary of the Armed Forces National Research Counsel Vision Committee until the mid-1950s. Blackwell was an associate professor at the University of Michigan where he directed its Vision Research Laboratories and later became director of the Institute for Research in Vision at Ohio State University. He died in 1995.

In 1946, while working at Tiffany, Blackwell published what turned out to be a seminal paper in human vision studies.[15] "Contrast threshold of the human eye" was one of the first significant works to examine the limits of the sensitivity of the eye to extremely small increments of contrast under controlled laboratory conditions. Our ability to visually distinguish among

[15] *Journal of the Optical Society of America*, 36(11), 624–643.

objects in our field of view depends on differences in their luminances (or surface brightnesses), their colors, or both; disregarding color for the moment, Blackwell investigated human subjects' reactions to targets whose luminances were very slightly higher or lower than their surroundings, yielding positive or negative contrasts, respectively. His results revealed the minimum contrasts required for visual detection over a large range of target brightnesses and angular sizes.

Blackwell's work built on Weber's Law, named for the German physiologist Ernst Heinrich Weber (1795–1878) who first articulated the "just noticeable difference" concept. Weber argued that the size of the difference threshold—the least amount by which a stimulus intensity must be varied in order to produce a noticeable modification of the sensory experience—is inversely related to the magnitude of the initial stimulus over a very large range of possible values. In other words, more intense stimuli result in smaller increments of change that are noticeable to an observer. Weber's Law is expressed as

$$\frac{\Delta I}{I} = k \tag{7.1}$$

where ΔI is the difference threshold, I is the intensity of the stimulus, and k is a constant.

Consider presenting an observer with two spots of light on a screen, each of which has an intensity of 100 in some arbitrary units. You give the observer a variable switch that allows the brightness of one of the spots to be changed at will by simply turning a dial, and you ask him or her to dial the brightness of that spot until it is just barely (but noticeably) brighter than the other spot. Let's say that number is 105 in the same units, or 5% brighter than the first spot. Weber's Law gives a k constant value for that observer of $5 \div 100 = 0.05$. Knowing that constant, one could predict the magnitude of the observer's difference threshold for light spots of any brightness provided that they were not excessively dim or bright. For example, if the brightness of the first spot were raised to, say, 1000 units, one could confidently predict that the observer's difference threshold in that case would be $0.05 \times 1000 = 50$. The observer would therefore need to vary the brightness of the second spot by at least 50 in order to confidently detect a difference in the brightness of the two spots.

Although Weber's Law for vision holds over a remarkably large range of target brightnesses, it doesn't tell the entire story. Again, setting aside color for the moment, some useful results are gathered from isolating out just the eye's rod cells, the photoreceptors most sensitive to faint light, which themselves offer no color discrimination to the brain. Repeating the kind of experiment

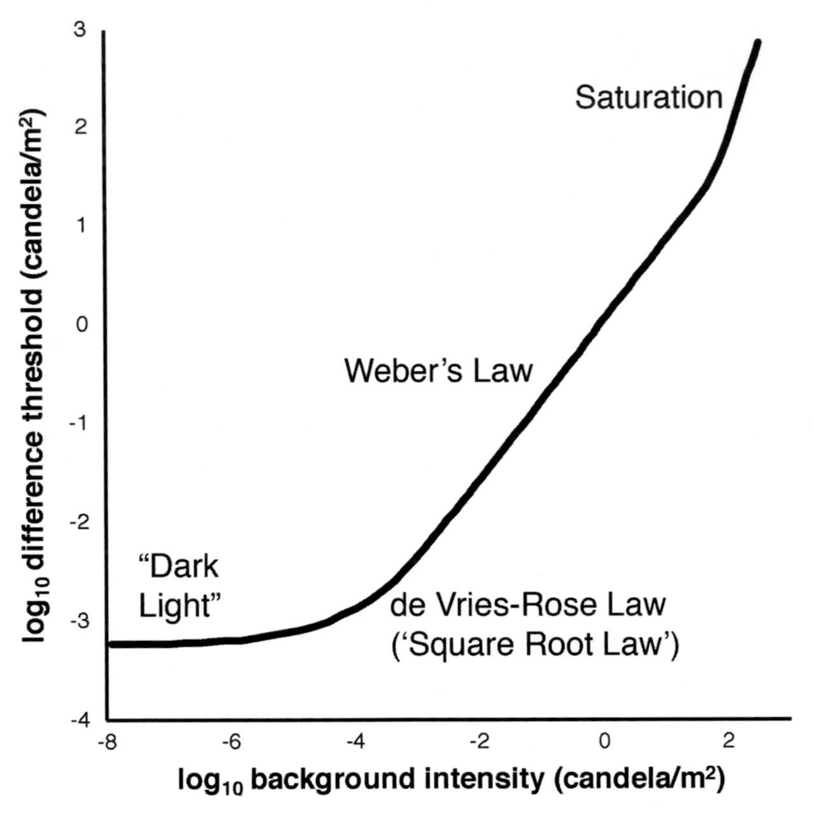

Fig. 7.5 Difference threshold for detection of objects against backgrounds of various intensities in vision dominated by rod cells. Adapted from Aguilar and Stiles, 1954, *Optica Acta*, 1, 59–65

in setting up Weber's Law over a very large range of brightnesses fills in some important details. Figure 7.5 gives the results in what's called a "log-log plot," both axes of which are shown in base-10 logarithms in order to compress a large range in both variables onto a single graph. It tracks changes in the difference threshold (vertical axis) as the brightness of the background next to the target object changes (horizontal axis).

The picture naturally divides into four regions. At the lowest brightnesses, "noise" in the visual system, set by biological processes, overwhelms the faint signal. It sets a floor beneath which the eye and brain do not perceive any further decrease in brightness. As the brightness slowly begins to increase, the threshold of detection switches over from the biological limit of the human visual system to the physical nature of the illuminated object and background. In this range, where the curve begins to turn up, the detection threshold rises roughly as the square root of the background intensity. Here the eye is presumed to operate like an ideal light detector, absorbing and registering each individual photon, or quantum of light. It is then particularly sensitive to statistical fluctuations in light emissions from the source, an effect which is seen in real (artificial) light detectors and often referred to as "shot noise." This region of the graph is called the de Vries-Rose Law. As the background level continues to rise, eventually the eye behaves according to Weber's Law, represented by the long straight-line part of the graph. Finally, at very high light levels, the rod cells begin to saturate, and the slope of the line quickly becomes more steep. At this point the rods are unable to detect changes in the stimulus.

Through his work at Tiffany in the 1940s, Blackwell filled in a lot of what we know about the Weber's Law regime of the graph. He constructed a special structure for the experiments consisting of a long, narrow room whose interior was painted with a high-reflectivity white paint. At one end of the room was a series of light projectors illuminating a screen at the other end, where an apparatus presented a series of targets. In an experimental protocol that would raise more than a few eyebrows today, Blackwell recruited as test subjects "young women, aged 19–26 years," whose uncorrected eyesight was at least 20/20 in both eyes. The reality of the situation was that recruiting a complement of similarly aged young men during the Second World War would have been difficult at best, although Blackwell has been since criticized for using only one gender of study participant.

The young women were seated in chairs along two rows at the back of the laboratory room adjacent to the light projector. Blackwell described the procedure:

> Stimuli, circular in form and brighter than the observation screen, were presented in any of eight possible positions on the screen for an exposure of 6 seconds…An electrical buzzer served as the signal for a stimulus presentation, remaining activated continuously during the presentation. Breaking the buzzer circuit served as a signal for the observers to indicate whether they had detected the presence of the stimulus. The duration of the stimulus varied among experiments. In each case, a 6 second rest period followed each stimulus presentation. Additional rests were scheduled so that the observers did not become excessively fatigued.

As a result of generous program funding, Blackwell was able to employ the subjects full time through the two-and-a-half-year course of the study. Each session consisted of the presentation of 320 different configurations of target and background brightness and the apparent angular size of the target;

by the end of the study, Blackwell and his assistants had gathered over 2 million individual evaluations of the stimuli, of which nearly half a million were considered of sufficiently high quality to be included in his analysis.

The data affirmed Weber's Law over a large range of brightnesses and provided the first careful, empirical measurements of critical contrast threshold with enough statistical power to draw strong inferences. In 1953's *Vision through the Atmosphere*, the Canadian atmospheric scientist W. E. K. Middleton (1902–1998) praised the work. "The writer believes that there is every reason to accept the Tiffany data," Middleton wrote, "within the limitation that they are obtained exclusively by young observers, and the probable further limitation that there should be no boundaries of much higher contrast in the immediate vicinity of the object to be detected."

In a certain sense, the degree to which the eye is sensitive to small changes of contrast seemed to have been determined rather definitively by Blackwell in the Tiffany experiments. But that determination was made in carefully controlled laboratory conditions and at relatively high light intensities. What happens when the eye is in real-world conditions in the field, particularly at very low light levels? And how would that knowledge bring us closer to understanding how the eye perceives the Ashen Light through the eyepiece of a telescope?

In 2014, Andrew Crumey of Northumbria University in Newcastle, England, published a paper titled "Human Contrast Threshold and Astronomical Visibility" in which he brought together the work of Blackwell and others and applied it to finding the limits of human eyes viewing the night sky, both with and without optical aid. Crumey began his paper with a straightforward statement of its motivation:

> Determining the faintest star or extended object visible to the naked eye or with a telescope is a problem of interest in light pollution studies, the history of astronomy, and vision science. ...This article will present a model applicable to uniform achromatic targets of any size,[16] seen against background luminance levels ranging from total darkness to daylight, hence relevant to visibility problems in many areas. For low light levels it will be applied to historical astronomical data and shown to be more accurate than previous models.

Crumey developed a visibility model for faint astronomical phenomena that seems to have the most relevance to the problem of the Ashen Light in terms

[16]In this context, 'achromatic' means that the targets are not very strongly colored and are therefore treated as though they were essentially white. The colors of such objects are perceived to have zero saturation and therefore no hue.

of figuring out what threshold brightness relative to the surrounding night sky is sufficiently large to trigger detection by a telescopic visual observer of Venus. It became clear that Crumey's model could be the bridge between historical observations and the airglow mechanism that seemed to stand the best chance of physically explaining the Ashen Light.

I sent a message to Crumey congratulating him on the paper and describing what I saw as the application to the Ashen Light while asking, in so many words, *is my thinking correct here?* He was intrigued and quickly replied to me. He recognized the parallel between the faintly illuminated night side of Venus seen against a (darker) background and the controlled illumination of targets by Blackwell in his experimental theater. Crumey cautioned, however, that it's not always appropriate to assume that experimental conditions in the laboratory match viewing in the real world especially well. "Blackwell dealt with uniform discs against uniform backgrounds," Crumey wrote. "What we would have with an ashen-lit Venus would be a divided disc with two levels of illumination, against a uniform background." The shape of the bright and dark parts of the Venus disc would be an important complication that his model didn't address.

But what about glare from the part of Venus directly illuminated by sunlight? His model didn't account for such circumstances. "Suppose we were to imagine that the target (the unilluminated part of Venus) is modeled as a uniform disc of the same angular area," Crumey wrote, "and the illuminated part is modeled as a disc (or point source) of the same illuminance as the real surface. We would then treat the illuminated part as a glare source." In the late 1980s, a professor at the University of Waterloo School of Optometry, Werner K. Adrian (1930–2008), dealt simply with the issue of glare by treating it as an additive component that would have the tendency of raising detection thresholds. That approach matches the anecdotal experience of observers in the historical record, who found the Ashen Light much easier to see when the bright Venus crescent was hidden behind a field stop or occulting bar placed near the focus of the eyepiece.

Whether or not the bright crescent is visible also makes an important difference in terms of which visual receptor system–rods, cones, or both–is active and therefore to which wavelengths of light the eye is most sensitive at the time of observation. Subject to details, the crescent could well be sufficiently bright to trigger the cones and color vision, but it would come at the expense of sensitivity to the considerably lower light levels expected for the Ashen Light if it were explained by airglow in the Venus atmosphere. The eye in cone-dominated vision would also suffer from a lack of sensitivity at the wavelength of the green oxygen line, whereas the dark-adapted eye dominated by the rod cells would be considerably more likely to sense this light. In fact, failure to account for the bright crescent might rule out detection of faint Ashen Light altogether.

But Crumey's visibility model can handle this. "In my applications of the model," he wrote to me,

> I assumed scotopic vision, however I would suspect that telescopic observations of Venus could employ mesopic vision. As long as the target and background can be considered achromatic (which they should be) it ought to be possible to apply my model, but rather than the scotopic model it would need to be the photopic one.

"Mesopic" vision is an amalgam of the two modes taking into account bright (photopic) and faint (scotopic) light; it combines, in a limited sense, the color discrimination of cone-dominated vision with some of the faint-light sensitivity of rod-dominated vision. Most often, we experience mesopic vision in outdoor contexts after dark in urban and suburban situations where the light levels never become quite low enough for the rod cells to completely take over.

Careful and well-prepared observers used field stops and occulting bars to keep the glare of direct light from the sunlit side of Venus out of their eyes; coupled with mechanical or motor drives that compensate for the rotation of the Earth by keeping objects motionless in the telescope's field of view, it's entirely possible that some eyewitnesses of the Ashen Light were fully dark adapted when they made their observations. On the other hand, some historical accounts indicate that observers were quite unprepared for what they saw. This is true going all the way back to the first report by Riccioli in 1643, who described the illuminated portion of the Venus disc as so bright that it was fringed by bright colors induced by chromatic aberration in his uncorrected telescope lens. Clearly it's not enough to assume that observers were dark adapted when they saw what they identified as the Ashen Light. Even mesopically adapted eyes would need more light to make a confident detection, upping the ante in terms of how bright we must assume the Ashen Light is if real.

If one rolls all of this into Crumey's model, what comes out? For one thing, a remarkably low intensity of light seems to be needed to trigger detection by an observer on Earth equipped with a modestly sized telescope. Under optimal conditions, it may be possible for a visual observer to detect an oxygen green-line glow on Venus with an intensity of as little as about 10 Rayleighs (R). Tom Slanger's 2001 observations of the Venus airglow yielded an intensity of about 150 R, but in a 2006 follow-up study he and his colleagues remarked on its "substantially varying intensity" in observations made through 2004. On some nights, the green line was clearly below their detection threshold of about 10 R.

Putting this together with Candace Gray's strong detection of the green line on Venus during the Bastille Day Event, it began to add up logically. The episodic nature of the Ashen Light in historical reports and its higher likelihood of being seen around the time of solar maximum are consistent with a phenomenon like airglow that is triggered by the direct impact of solar charged particles onto the ionosphere of Venus. CMEs yield the brightest green-line glows, but they only affect the planet for a few hours. But the Ashen-Light-as-oxygen-green-line hypothesis also has to account for instances where the Ashen Light has been reported over multiple nights adjacent in time, particularly close to solar minimum. In such cases, CIRs as the source of energetic particles to excite oxygen atoms make more sense. Although they don't raise the plasma densities around Venus to the same extent that the passage of a CME does, the influence of CIRs on Venus lasts for days rather than mere hours.

We have a plausible mechanism, we have a visibility model that establishes a remarkably low detection threshold, and we have the weight of eyewitness descriptions of the Ashen Light over nearly four centuries. Is that enough to consider this case closed? Before sending the jury off to deliberate, we must give full and fair consideration to the alternative: that there is nothing objectively real about the Ashen Light at all and that every instance in which it was reportedly seen resulted from a cruel conspiracy of marginal conditions and a human evolutionary inheritance that tries to convince us that the wrong answer is in fact right.

8

Seeing What We Want to See: The Psychology of the Ashen Light

The fabled "canals" of Mars (Chap. 1), the claim for whose existence is traditionally ascribed to a single mistranslated Italian word and the global deference to one well-respected observer's insistence on what he saw through the telescope eyepiece, is hardly the only instance of mass delusion in the history of astronomy. Nor is it the only case in which spurious sightings drew astronomers to conclusions, later proven wrong, that were for some length of time taken as articles of faith. One such example involved a mathematical discrepancy, undiscovered physics, and a healthy dose of wishful thinking.

The Discovery of Neptune

On the night of March 13, 1781, the Anglo-German astronomer William Herschel was at the eyepiece of a small reflecting telescope whose primary mirror had a diameter of just over six inches. From the back garden of his house in New Kings Street, Bath, Herschel had begun his "second review of the heavens" in late 1779, using the telescope to make slow, careful sweeps of the night sky, and recording carefully all of the objects he noted as they wandered through the field of view. The positions of the stars interested him little, as others had already produced reasonably accurate star maps by his time. Rather, he was interested in what the maps *didn't* show. Working together with his sister, Caroline Herschel (1750–1848)—later recognized as an important astronomer in her own right—it was through systematic searching and diligent

J. C. Barentine, *Mystery of the Ashen Light of Venus*, Astronomers' Universe,
https://doi.org/10.1007/978-3-030-72715-4_8

record-keeping that he slowly built up a reputation as one of the best observers of his day.

While looking for and noting the position and nature of multiple star systems on that late winter evening, he stumbled upon a very curious object. "In the quartile near ζ Tauri," he wrote in his journal, "[is] either [a] Nebulous star or perhaps a comet."[1] Whatever the object was, its appearance did not resemble those of the stars, whose apparent sizes are seemingly invariant regardless of the degree of magnification to which they are subjected. Herschel later recounted his magnification tests that yielded no further useful information about the object:

> From experience I knew that the diameters of the fixed stars are not proportionally magnified with higher power, as the planets are; therefore I now put on the powers of 460 and 932, and found the diameter of the comet increased in proportion to the power, as it ought to be, on a supposition of its not being a fixed star, while the diameters of the stars to which I compared it were not increased in the same ratio.[2]

The compact object "appeared hazy and ill-defined with these great powers," leading Herschel to conclude that it was likely a previously unknown comet:

> On Tuesday the 13th of March, between ten and eleven in the evening, while I was examining the small stars in the neighbourhood of H Geminorum, I perceived one that appeared visibly larger than the rest: being struck with its uncommon magnitude, I compared it to H Geminorum and the small star in the quartile between Auriga and Gemini, and finding it so much larger than either of them, suspected it to be a comet.[3]

Herschel returned to the same region of sky on successive nights, finding the object drifting slowly toward the east, a pattern that could well indicate a comet still far from the Sun and just beginning to shine. Its attributes were not strictly cometary; for one thing, it showed no suggestion of a tail, but at that point it might simply have been too distant to yet have one. He wrote to his old friend Nevil Maskelyne (1732–1811), the Astronomer Royal, with the details; on April 23, a flustered Maskelyne replied:

---

[1] Royal Astronomical Society MSS W.2/1.2, 23, quoted in Miner, E., 1990, *Uranus: The Planet, Rings and Satellites*, New York: Ellis Horwood and Prentice-Hall, 8.

[2] 1781, *Philosophical Transactions of the Royal Society of London*, 71, 492–501.

[3] *Ibid.*

I don't know what to call it. It is as likely to be a regular planet moving in an orbit nearly circular to the sun as a Comet moving in a very eccentric ellipsis. I have not yet seen any coma or tail to it."[4]

In short order, the Russian astronomer Anders Lexell (1740–1784) computed an orbit for the new object showing it to be a low-eccentricity ellipse and therefore having much more in common with the planets than with short-period comets. More importantly, Lexell showed that the object was considerably more distant than Saturn. At such large distances, comets are yet too cold to shed considerable amounts of ice and dust to reflect sunlight and thus are generally invisible to small telescopes on Earth.

Herschel had become the first human since antiquity to discover a planet. Maskelyne asked Herschel to "do the astronomical world the faver [sic] to give a name to your planet, which is entirely your own, [and] which we are so much obliged to you for the discovery of."[5] The canny Herschel hoped to use the discovery to snag the attention of King George III, who he figured might well reward the first discovery of a planet from British soil with a little royal largesse. He therefore dubbed the new planetary interloper "Georgium Sidus" (George's Star):

> In the fabulous ages of ancient times the appellations of Mercury, Venus, Mars, Jupiter and Saturn were given to the Planets, as being the names of their principal heroes and divinities. In the present more philosophical era it would hardly be allowable to have recourse to the same method and call it Juno, Pallas, Apollo or Minerva, for a name to our new heavenly body. The first consideration of any particular event, or remarkable incident, seems to be its chronology: if in any future age it should be asked, when this last-found Planet was discovered? It would be a very satisfactory answer to say, 'In the reign of King George the Third'.[6]

International disagreement concerning the name, including much consternation from the other side of the English Channel that a new planet would be named after a foreign monarch who still nominally claimed the throne of France, ensured that Herschel's suggestion fell out of favor. The winning name was ultimately put forth by the German astronomer Johann Elert Bode (1747–1826): Uranus, the father of Saturn in classical mythology. Because Saturn

[4]Royal Astronomical Society manuscript Herschel W1/13.M, 14, quoted in Miner, 1990, 8.

[5]Royal Astronomical Society manuscript Herschel W.1/12.M, 20, quoted in Miner, 1990, 12.

[6]Letter to Joseph Banks, quoted in J. L. E. Dreyer, 1912, *The Scientific Papers of Sir William Herschel*, London: The Royal Society, Vol. 1, 100.

was the father of Jupiter, Bode argued, the name "Uranus" was both highly appropriate and nationally neutral. The name stuck.

Over the following decades, additional observations were used to compute what was believed to be a precise orbit for the new planet. And while the new planet itself was a startling discovery, its observed motion was not. It adhered beautifully to the prescriptions of Isaac Newton (1642–1726), whose universal theory of gravitation was published just under a century before Uranus wandered into the view of Herschel's telescope. Since the time of Newton, scientists held out hope that his theory would describe a kind of clockwork perfection for the motion of objects in the Solar System. If one completely accounted for the forces acting on a system, Newton's theory held, then the future evolution of the system's motion could be predicted with absolute accuracy. But although Newton trusted the mathematics and physics on which his theory was based, he doubted the reliability of its forecasts over long periods of time. He thought that maintaining the order of the Solar System required periodic corrections and that only God was capable of guaranteeing the enduring harmony of the cosmos.

Two generations after Newton, the French polymath Pierre-Simon Laplace (1749–1827) set himself the task of proving the perfection of Newtonian theory and dispensing with the notion of divine influence over the physical universe once and for all. Laplace started with the assumption that such perfection was possible a priori, unaware of the inherent unpredictability of complex mechanical systems and their tendency toward chaotic behavior. He also had no way of knowing, as we'll soon find out, that Newton's ideas were not the last word on the human understanding of gravity.

An acute problem in observational astronomy during Laplace's time had to do with the relative sizes of the orbits of Jupiter and Saturn, namely, that the former appeared to be contracting while the latter was expanding. A century and a half of observations since the invention of the astronomical telescope yielded positional data of sufficient precision that it was possible to detect tiny deviations of the planets' positions from the predictions of Newtonian theory. The measurements suggested that something about the outermost planets' orbits was changing. Newton would have insisted that only gravity could account for this situation and that therefore any errors resulted from an incomplete accounting for the forces acting on the planets in their orbits. The greatest mathematical minds of the eighteenth century tried solving the problem without success. Laplace considered solutions in which something pervading the cosmos was exerting a drag force on the planets and slowing them in their orbits, as well as the possibility that the influence of gravity might not be instantaneously transmitted from one body to another as Newton's theory implied.

Laplace pursued these ideas into the 1770s before deciding that they were inadequate, which strengthened his conviction that Newton's laws of motion and gravitation were fundamentally correct. Earlier approaches to the problem tended to ignore small, high-order terms in the equations of motion of the planets; Laplace guessed correctly that small terms had a large cumulative effect when integrated over long periods of time. Actively including these terms in the calculation of the planets' motion was laborious, but it did improve the accuracy of their predicted positions. Laplace's conclusion was simple: it was only humans' relative inability to make all of these tedious calculations that explained any discrepancies between the predicted and observed positions of the planets and not some fatal flaw in Newton's theory that could only be overcome by a periodic push from the hand of the Almighty. From this he decided that all objects in the Solar System must be in mutual gravitational equilibrium with each other, their motions carefully prescribed by Newton's laws only.

By the early nineteenth century, having been rigorously tested, belief in the reliability of Newtonian theory was strong. The crowning glory of this belief was Laplace's *Traité de mécanique céleste*, published in five volumes from 1798 to 1825. At the time of Laplace's death in 1827, many were convinced that he really had solved one of the great physical and mathematical problems of the ages. To the precision of observations available in his time, he was right; improved data in the future would require only more sophisticated calculations and not further corrections to the underpinnings of physical intuition about how the universe worked. If these calculations were carried out reliably and fed with more observations in an iterative fashion, they would yield dependable planetary positions into an indefinite future. Even the discoveries of new Solar System objects, such as the asteroids, were thoroughly accounted for by Newtonian physics and Laplacian mathematics. There was good reason to believe this was the end-all and be-all of planetary physics. Man had read the mind of God, revealed in creation, and copied down his thoughts with complete authenticity.

So when discrepancies in the position of Uranus with respect to mathematical predictions began to creep into observational data in the first decades of the nineteenth century, faith in Newton was initially unshaken. In 1821 the French astronomer Alexis Bouvard (1767–1843) published tables containing future positions of Uranus on the night sky using only Newtonian theory and the combined gravitational influence of the known planets. Inconsistencies noted by observers were just small enough to dismiss as the result of careless record-keeping. Tellingly, deviations existed not only in the position of Uranus along the apparent path of the planets across the sky but also in the inferred length and direction of the invisible line connecting the centers of Uranus and the Sun. Something was not only influencing the planet's motion perpendicular to our line of sight but also along it. Astronomers began to wonder whether something was periodically pulling Uranus in its orbit, causing it to speed up or slow down. But was that unseen influence the hand of God or something decidedly less divine?

Despite decades of observations, the irregularities didn't go away. More position measurements confirming the deviations made it seem less likely that they were the result of clerical errors. Maybe gravity somehow worked differently over very large distances such as that between Uranus and the Sun. But no one ever showed any evidence that the apparent motion of Saturn wasn't completely accounted for, within observational uncertainties, when the gravitational leverage of both Jupiter and Uranus was properly considered.

Taking the Jupiter-Saturn-Uranus analogy a step further, perhaps an unseen object existed that provided a kick every time Uranus overtook it in its orbit and that would explain errors in the predicted positions of Uranus. Bouvard suggested this was most likely because all of the other known planets obeyed Newton so strictly. Given that nothing like another Uranus ever turned up in the many sky-sweeps of Herschel and other observers, the perturbing planet must be either very faint, moving extremely slowly across the sky, or both. The problem began to resemble that of finding the proverbial needle in a cartoonishly large haystack. Over a generation before the then-new process of photography would be applied to astronomical observations, how could one even begin a search for a heretofore unseen planet with the slightest probability of success? Two men, one English and the other French, independently decided to find out.

In the spring of 1843, John Couch Adams (1819–1892; Fig. 8.1) was finishing up his studies at St. John's College, Cambridge, when he read about the unsolved problem of Uranus's orbital irregularities. The eldest of seven children born to a poor tenant farmer in a small village near Launceston, Cornwall, Adams became fascinated at an early age by astronomy books among the collection in a small library that his mother inherited from her uncle. A bright boy, Adams was afforded a modest private education away from home in Devonport, Plymouth, at a school run by his mother's cousin, the Rev. John Couch Grylls. At Grylls's school he received a standard, classics-based education common in his social class of the era, but in his spare time, he taught himself mathematics. Inspired by the 1835 apparition of Halley's Comet, he began making his own calculations and predictions involving astronomical

Fig. 8.1 Contemporary images of John Couch Adams (left, photograph *c.* 1870) and Urbain Le Verrier (right, engraving from unknown date)

phenomena. While still a student himself, he started offering private tutoring as a means of financing his work.

The following year, his mother inherited a small estate. Adams was accepted to university; the family's windfall enabled him to take the offer, and he moved to Cambridge where he entered St. John's in October 1839. There he honed his mathematical skills to the point that he felt the Uranus problem was tractable in a reasonable period of time. After making some initial calculations he began to feel that Bouvard was right and that an object beyond Uranus was in fact responsible for the observed positional discrepancies. Adams convinced himself that by using nothing more than Newton's Laws and existing planetary position data, he could infer the mass, position, and orbital characteristics of the unseen object. Just after completing his final exams in 1843, he returned home to Cornwall and set to calculating the orbit of the planet he believed to exist based only on its gravitational influence.

The problem before Adams was to "invert" the positional data in order to figure out what was causing the observed motion of Uranus. It's easy enough to do in the twenty-first century through brute-force methods and heavy computing power, but inversion of observational data in the nineteenth century was extremely time-consuming, and tedious hand calculations were prone to error. Adams used an iterative method with initial conditions, calculating the anticipated position of Uranus with perturbations from the unseen planet and then taking the residuals between the calculated and observed positions. The magnitude and sign of the residuals gave him a sense of whether his orbit for the perturber under- or over-corrected its actual motion.

Through a process like a regression analysis, he made changes to the elements of the perturber's orbit until it minimized the residuals. But the landscape of the variables was largely unknown, so the strong possibility existed that Adams would fall into a local minimum far from the right answer. Only adding new observations to the analysis might lead him steadily in the right direction, so he kept feeding new data to his model in hopes that the result would point increasingly to a *global* minimum and the correct orbit for the new planet. With an orbital solution in hand, he could compute a position on the night sky for telescopic observers to carefully search.

Adams requested additional Uranus observations during 1844 from the Royal Observatory through the influence of James Challis (1803–1882), Director of the Cambridge Observatory. Challis relayed the request to the Astronomer Royal, George Biddell Airy (1801–1892), at Greenwich, who readily complied. By the autumn of 1845 Adams completed his calculations and felt that he could confidently say where in the night sky the planet would be found. Adams communicated the results to Challis sometime in late 1845 or early 1846, but didn't send any details on his method. Challis was not impressed. As a result, he was not encouraged to devote telescope time to a search program, remarking "it was so novel a thing to undertake observations simply in reliance upon theoretical deductions,—while the labour was certain, success appeared to be so uncertain,—that it is not surprising a disposition should have been felt to postpone such observations to others based on less speculative grounds."[7]

Unbeknownst to Adams, the competition on the other side of the English Channel was hard at work. Urbain Le Verrier (Fig. 8.1) was born at Saint-Lô, Manche, in 1811. His father was an estates manager and an official in the imperial government, providing his family with a comfortable living in the countryside. After completing his studies in Saint-Lô at the age of sixteen, he enrolled in the College Royal at Caen. Despite showing great promise, he failed the competitive exam for admission to the École Polytechnique in 1830. Le Verrier's father, Louis-Baptiste, was convinced of his son's talent and, like any good parent with social ladder-climbing ambitions, found that he could

[7]1846, *Memoirs of the Royal Astronomical Society*, 16–17, 416.

buy Urbain's way to higher education. Louis-Baptiste sold the family home in Saint-Lô and used the proceeds to send his son to the Mayer Institute in Paris.

After a year of study, the younger Le Verrier won second place in the Concours Général, the most prestigious academic competition in France, which entitled him to admission to the École Polytechnique. He graduated from the École in 2 years, after which he spent another two studying industrial chemistry at Orsay. Hoping for a career in the tobacco industry, he worked to understand the reactions of phosphorus with hydrogen and oxygen in order to determine the influence of phosphorus matches on the taste and perceived quality of tobacco. Le Verrier continued his work at Orsay under the direction of the chemist Louis Joseph Gay-Lussac (1778–1850), his professor at the École, who was already well-known for his work on the chemical composition of substances like water, and on the laws governing the physics of gases.

Offered a position teaching chemistry in the provinces in 1836, Le Verrier elected to stay in Paris and continue working with Gay-Lussac. However, like virtually all scientific research undertaken outside the system of private patronage common at the time, the work didn't pay. To earn a living, Le Verrier took a job as a part-time instructor at the Collège Stanislas and gave private lessons in mathematics. The same year, two *répétiteur* positions opened at the École Polytechnique. In the French academic system of the nineteenth century, *répétiteurs* were teaching assistants who conducted what we might now call recitation sections in which they questioned students on the subject matter of their professors' lectures. Le Verrier applied for the position that reported to Gay-Lussac, but he lost out to Victor Regnault (1810–1878), a fellow chemist who made a strong application on the basis of his organic chemistry work. Slightly wounded, Le Verrier applied for the second *répétiteur* position, reporting to the astronomer Félix Savary (1797–1841). Winning the position, which Le Verrier accepted, changed the trajectory of both his career as well as his life. He wrote to Louis-Baptiste:

> In daring to accept the duties which had successively been fulfilled by François Arago, Claude-Louis Mathieu, and Félix Savary, I have imposed on myself the obligation not to let the post which they have occupied be depressed in the public esteem, and for this I must not only accept but seek out opportunities to extend my knowledge. … I have already ascended many ranks, why should I not continue to rise further?[8]

---

[8]Lequeux, J. 2013. *Le Verrier—Magnificent and Detestable Astronomer*, New York: Springer, 4.

Le Verrier's first paper in the field, "Sur les variations séculaires des orbites des planètes," presented to the Académie des Sciences in September 1839, led him directly to the problem of the stability of orbits in the Solar System. He then made a study of the orbits of periodic comets, showing that historical accounts of certain comets were really unknowing references to the same objects perturbed into new orbits by the gravitational influence of Jupiter. He became well-known in the astronomical world and by the mid-1840s was elected to the Académie.

Around the same time, he began working on the problem of Uranus's irregular motions, entirely unaware of Adams's effort, and yet coming to the same conclusion that an unseen planet must be the cause. On November 10, 1845, he presented his idea to the Académie in the first of a series of *mémoirs*, further refining his calculations through the winter and spring of 1846. By early summer, in a second *mémoir*, he ventured a guess at the undiscovered object's position on the night sky. But he had yet to arrive at results sufficiently robust to conclude anything about its mass or orbital characteristics.

Meanwhile, back in England, Airy read the public account of Le Verrier's second presentation in late June and quickly recognized the similarities in the predictions made by Le Verrier and Adams. This prompted him to try to scoop the rest of the astronomical community by mounting a secret search program at Greenwich to find the new planet before others even began to plan their reconnaissance of the night sky. Airy hastily arranged a meeting of the Royal Observatory Board of Visitors, with Challis and John Herschel (1792–1871), son of William, in attendance, at which he suggested that a search using the Cambridge 11.25-inch refractor should commence at once. He pressed the Visitors on the importance of his work, "in the hope of rescuing the matter from a state which is ... almost desperate."[9] The search began in earnest on the night of July 29.

Throughout the summer of 1846, Adams continued iterating his solution for the orbit of the Uranus perturber, offering the English search team a half-dozen new position predictions. The last of this series, made on September 2, gave a position 315.3° east of the vernal equinox along the ecliptic, which was in error by 12°. Despite his best efforts, Adams managed to send Challis and his assistants to the wrong part of the night sky, far west of the perturber's actual position. The search method Challis employed involved wide sweeps along the ecliptic, and purely by chance he scanned its actual position twice during

[9] Smart, W. M. 1989. John Couch Adams and the discovery of Neptune, *Occasional Notes of the Royal Astronomical Society*, 2, 59.

August. But, lacking the latest and most accurate celestial charts, he failed to recognize the interloper wandering slowly among the background stars.

Given the secrecy of the program, neither Le Verrier nor anyone else on the Continent knew anything about it. On August 31, Le Verrier was back at the Académie presenting another *mémoir*. His calculations had advanced to the point where he was able to specify the mass and orbital elements of the unseen planet, but to this point in time, no one in France with access to a suitable telescope seemed interested in searching for it. Seeking a friendly collaborator with that access, he sent a letter to Johann Gottfried Galle (1812–1910) at the Berlin Observatory. Galle was awarded his doctorate in 1845 for a thesis in which he performed a careful and precise reduction of meridian transits of various stars and planets obtained by the Danish astronomer Ole RØmer in 1706. Galle sent a copy of his thesis to Le Verrier, who waited a year to respond and then only did so apparently when he needed something from him.

Nevertheless, interested in the prospect of participating in a major astronomical discovery, Galle immediately began a search around Le Verrier's predicted position on the night of September 23, mere hours after receiving his letter. Permission to use telescope time for the search was granted by the director of the Berlin Observatory, Johann Franz Encke (1791–1865), who reluctantly agreed despite his own unease with the reliability of Le Verrier's prediction. Galle had one chance to get it right, as Encke was unlikely to afford him unlimited time to look for something that might well be nowhere near where Le Verrier said it would be found.

Galle's student, Heinrich Louis d'Arrest (1822–1875), suggested that the pair consult the latest and most accurate star charts in the region around the predicted location of the new planet. It was expected that a more distant object than Uranus would be proportionately dimmer and thus would look even more like a background star than Uranus did when William Herschel first spotted it some 65 years earlier. Any "star" within the range of brightness plotted on the chart that looked out of place was a suspect, but confirmation of the detection would require at least a night or two in order for the object's motion against the actual stars to betray its true nature.

Luckily for Galle, he spotted something just past midnight, after a search of less than an hour. Comparison of the field with the *Berliner Akademischen Sternkarte* sky atlas showed nothing of comparable brightness at the location, which was less than 1° away from Le Verrier's predicted position. Galle and d'Arrest had no other option but to contain their excitement; at the expected distance of the new planet, its orbital drift against the stars was too small to notice in only a few hours of observing.

They came back on the following two nights, during which they were blessed with sufficiently clear skies to check the area through the telescope and gather updated information on the object's position, direction, and rate of motion. They found a rate of 4 seconds of arc per day, again matching expectations based on the presumed orbit of the Uranus perturber and the predictions of both Le Verrier's and Adams's models. The following day, Galle wrote to Le Verrier with the news: "Monsieur, the planet whose place you have [computed] really exists."[10] The news spread quickly across Continental Europe. Arago initially suggested the name "Le Verrier," but, following the example of Uranus, Le Verrier proposed the name "Neptune." Galle himself later refused credit for the discovery of Neptune, attributing it solely to Le Verrier's computational prowess.

After the discovery was announced, Challis, John Herschel, and Richard Sheepshanks, foreign secretary of the Royal Astronomical Society, publicly claimed that Adams had beaten Le Verrier to the punch and predicted the correct position of Neptune first. Airy wrote his own version of the story praising Adams over the rival Frenchman, but he left out crucial details of the story that undercut the English claim. For instance, he failed to mention that Adams only quoted a mean ecliptic longitude for the object and not its orbital elements, which made it impossible for others to check his work by computing their own position predictions. Airy found himself on the receiving end of much criticism in England for his account once the details of Le Verrier's work made their way across the Channel.

Although Adams's method was shaky, his contribution to the pre-discovery prediction of Neptune's existence might have been better secured for posterity had he the self-confidence as a young researcher to believe in his own results. Responsibility for his failure to promote himself and his work was laid at the feet of his mentors, Airy and Challis, who themselves became targets for charges that they failed in their roles. Challis eventually grew contrite about his part in the affair, although Airy tried to shake off some of the negative judgment with a claim, now seen as patently ridiculous given his attempt at subterfuge in the summer of 1846, that it was never properly the responsibility of the Royal Observatory to undertake a search for Neptune.

France saw the discovery as Le Verrier's alone, with no substantial aid by Adams, and ensured its native son's fame. Eventually even the Royal Society recognized the achievement, awarding the Copley Medal to Le Verrier later in

[10]Yanofsky, N.S. 2013. *The Outer Limits of Reason: What Science, Mathematics, and Logic Cannot Tell Us*, Cambridge/London: MIT Press, 306.

1846, "for his investigations relative to the disturbances of Uranus by which he proved the existence and predicted the place of the new Planet; the Council considering such prediction confirmed as it was by the immediate discovery of the Planet to be one of the proudest triumphs of modern analysis applied to the Newtonian Theory of Gravitation."[11] Proving that there were no hard feelings, and yet clearly laying claim to a key part in the story, Adams smoothed over the controversy by publicly acknowledging the priority of Le Verrier's claim (and Galle's observations) in a paper given to the Royal Astronomical Society in November 1846:

> I mention these dates merely to show that my results were arrived at independently, and previously to the publication of those of M. Le Verrier, and not with the intention of interfering with his just claims to the honours of the discovery; for there is no doubt that his researches were first published to the world, and led to the actual discovery of the planet by Dr. Galle, so that the facts stated above cannot detract, in the slightest degree, from the credit due to M. Le Verrier.[12]

In contrast, Le Verrier was arrogant about his role in the affair. Despite the fact that the discoveries of Uranus and Neptune gave England and France one new planet apiece, their respective scientific ranks held fast to their own countrymen as the responsible parties for the find in the autumn of 1846. But a gentleman to the end, Adams acted graciously toward Le Verrier years later when, as President of the Royal Astronomical Society, the task of presenting the RAS Gold Medal to Le Verrier fell to him.

Although the prediction of Neptune's existence and position with sufficient detail to enable its discovery was a successful test of Newtonian theory, the modern view places the achievement in appropriate context. Both Adams and Le Verrier were unduly confident in the precision of their calculations, and each grossly overestimated the mean distance of Neptune from the Sun. They arrived at an expected distance through the application of the Titius-Bode "law," a semi-empirical hypothesis about the spacing of the planets' orbits about the Sun that has no real physical basis. If the mean distance between the Earth and the Sun is defined as one "astronomical unit" (AU), Titius-Bode predicts an orbital distance for Neptune of about 38 AU. Manifesting their faith in the law's predictions, which found Uranus within 2% of its

[11]While it made no mention of Adams, in 1848 the Royal Society awarded him the Copley Medal in his own right "for his investigations relative to the disturbances of Uranus, and for his application of the inverse problem of perturbations thereto."

[12]1851, On the Perturbations of Uranus, *Appendices to various nautical almanacs between the years 1834 and 1854*, Nautical Almanac Office, London: W. Clowes and Sons, 265–293.

expected distance from the Sun, both Le Verrier and Adams started with the presumption that Neptune must be at the predicted distance.

Furthermore, as André Danjon pointed out in an extensive analysis on the centennial of Neptune's discovery,[13] the whole business might have been an accident attributable to the part of its orbit in which Neptune happened to be in 1846. Danjon showed that the orbits computed by Le Verrier and Adams, while reasonably alike, were remarkably far from the orbit later determined on the basis of many years of observations. However, by apparent chance, they were both closer to the actual orbit in the 1840s than for other periods of time both before and after the discovery. This had the fortuitous effect of offsetting the fundamental error in the mean distance from the Sun, diminishing its importance as a relevant orbital parameter.

Nevertheless, it's clear that the success of Newton's ideas in other realms inspired sufficient confidence in the celestial mechanics of the nineteenth century to suggest a rational approach to solving the problem of apparent inconsistencies in the orbit of Neptune. Over a century after Newton's death, astronomers no longer felt the need to conjure the occasional, arbitrary push of the divine hand to keep the Solar System in line—or to explain why sometimes things didn't behave exactly according to expectations. So when similar headaches arose resulting from observations of other known planets, researchers knew what tools to bring to bear on solving the problem, tools that had been validated in spectacular fashion beyond the orbits of the known planets. But the next application of Newtonian mechanics to the planets would only show the shortcomings of this framework, leaving two generations of astronomers puzzled.

The Cautionary Tale of 'Vulcan'

Before Urbain Le Verrier took up the question of what was pushing Uranus around in its orbit, he was approached about a similar state of affairs much closer to home. In 1840, François Arago, the Director of the Paris Observatory, suggested to Le Verrier that he look into historical records of the position of Mercury with an eye toward deriving a rigorous solution for its orbit around the Sun. We have seen already that Le Verrier was inclined to the topic during his early years as an astronomer, so he went away for a time and delved into what was then known about the motion of Mercury.

[13] 1946, *Ciel et Terre*, 62, 369.

Being an "inferior" planet whose orbit keeps it at all times nearer to the Sun than Earth, and owing to its rapid 88-day orbital period, the first of the terrestrial planets never strays far in the sky from the Sun itself. At most it appears no more than 28° east or west of the Sun, severely limiting the times when astronomers might study it in any conditions except full daylight; consequently, making highly precise positional measurements of Mercury is tricky.

While its orbit was reasonably well known, Le Verrier wanted to derive a full, formal solution for the orbit from first (Newtonian) principles. In 1843 he published "Recherches sur l'orbite de Mercure et sur ses perturbations,"[14] a tentative theory describing Mercury's motion that could be tested during an event called a transit. Happening in intervals of 3 to 13 years between successive events, the planet takes an excursion perpendicular to the direct line of sight between the Earth and Sun. Over the course of several hours, Mercury is seen in silhouette against the Sun's bright disc, a small and perfectly round dot that appears to slowly move relative to sunspots. Le Verrier realized that careful observations of Mercury transits would fill in important but otherwise missing gaps in humanity's knowledge of the planet's orbit, given the impossibility of otherwise observing Mercury at inferior conjunction when it is lost in the Sun's glare.

After the data from the 1843 Mercury transit were in, Le Verrier folded them into his model, turned the mathematical crank, and looked carefully at the results. The theory failed to correctly match the observations. What was going on? Again, it seemed like the departures of the observations from the theory were of a magnitude and direction such that something other than the gravitational influence of the Sun and the other (known) planets must be tugging on Mercury. Le Verrier mostly set the problem aside while working on what became the predicted discovery of Neptune, taking it up again in the mid-1850s. To rule out anything systematically weird about the 1843 transit observations, he collected and analyzed measurements of Mercury's motion during thirteen other transits between 1697 and 1848, along with scores of meridian timings, in which astronomers made very precise records of the instant that Mercury passed through the invisible line in the sky connecting the north and south celestial poles.

But residuals to his orbital solution for the planet remained that were not explained by Newtonian mechanics. Specifically, Le Verrier found that the location of the point at which Mercury reaches perihelion, the closest point in

---

[14] *Journal de mathématiques pures et appliquées 1re série*, 8, 273–359.

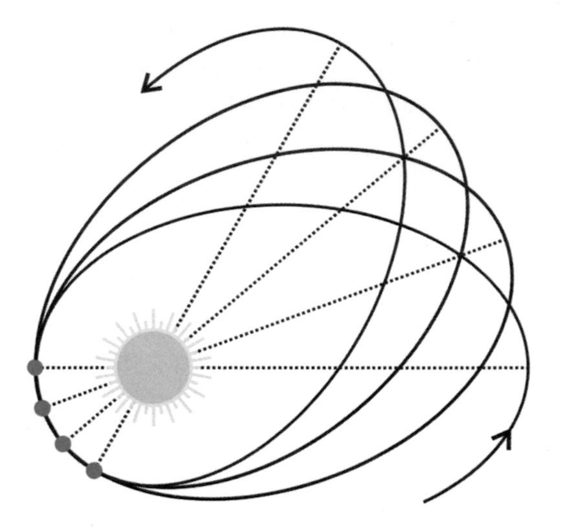

Fig. 8.2 Schematic drawing of a planet (blue circle) whose orbit around the Sun is precessing. With each lap completed, the orbit's major axis (dotted line) advances in angle

its orbit around the Sun, was slowly changing through time by an incredibly small but measurable amount—43 seconds of arc, or a little more than 0.01°, per century (Fig. 8.2). It appeared that the orbit of Mercury was precessing around the Sun in a way that could not be accounted for by the gravitational influence of the Sun and known planets alone, strengthening the case for a then-unknown perturber *interior* to Mercury's orbit.

Le Verrier published the contention in 1859,[15] igniting a new controversy: how could humanity have totally missed a planet large enough to pull on Mercury in a way that explained the observations? Admittedly a hypothetical innermost planet would never appear far from the Sun in the sky. Perhaps it was just not bright enough to be seen against all that glare? An "intra-Mercurian" planet with an orbit at all resembling those of the other planets would, like Mercury and Venus, periodically transit the Sun, but no claim of any observed transit was then known. Though such transits might be more frequent than those of Mercury, their duration might well be considerably shorter. Alternately, maybe people just weren't looking at exactly the right time to serendipitously notice something they had no prior reason to expect.

[15]Lettre de M. Le Verrier à M. Faye sur la théorie de Mercure et sur le mouvement du périhélie de cette planète, *Comptes rendus hebdomadaires des séances de l'Académie des sciences'* (Paris), 49, 379–383.

Skeptics devised various ad hoc fixes for the orbital solution, searching for anything that didn't involve the embarrassment of an undiscovered planet more or less right under Earth's proverbial nose. Could it be, for instance, there was a large amount of dust in the vicinity of Mercury's orbit? The planet's motion through a reasonably dense cloud or ring of dust would exert a drag force that slowed it down, mimicking the perturbing effect of a phantom planet. But this explanation, like the others, was knocked down when further observations failed to turn up the slightest bit of supporting evidence.

It turns out that the right answer was utterly unknown to nineteenth century scientists, although they had already begun to sense the outermost edges of the new physics that the answer represented. In the Newtonian view, a two-body system consisting of a lone object orbiting a large, perfectly spherical mass would trace out the shape of an ellipse with the large mass nearly stationary at one focus. The point of perihelion of the smaller object in this framework is fixed, and the object returns to the same point with every trip around the more massive object. Of course, the real Solar System is composed of more than one planet, and the gravity of all the others imparts a collective kick that perturbs the orbit of our test object. With some tedious calculations, one can account for these influences and make a first-order correction to the elements of the planet's orbit.

Another influence, albeit much smaller in magnitude, arises from the fact that the Sun *isn't* a perfect sphere. Since the Sun acts like a fluid body, it flattens slightly at the poles and bulges out slightly at the equator due to its rotation. This "oblateness" modifies the three-dimensional shape of the gravitational field of the Sun and results in a slightly different trajectory for each of the planets than would be the case if the Sun were a true sphere. This, too, can be accounted for in orbital calculations. There are a host of even smaller effects, each of which can also be modeled and subtracted out. But for Mercury, the departures from Newtonian predictions remain.

What Le Verrier didn't know, and humanity itself wouldn't understand for another half-century, is that Newton's laws weren't the end of the story in explaining how gravity works. The reason is that the Newtonian view of space and time was only appropriate for certain situations in which bodies with relatively small masses moved at relatively slow speeds. In the first years of the twentieth century, Albert Einstein (1879–1955) came to understand Newton's laws as a limiting case of a more general theory of gravitation. In 1915, Einstein proposed that humanity's picture of gravity was fundamentally incomplete. If applied to appropriately small masses and distances, Newton's theory worked pretty well. But move up to very large masses and increasingly larger distance scales, and the errors pile up quickly.

To account for this, Einstein devised the notion of "spacetime," a four-dimensional universe in which the three spatial coordinates (up-down, right-left, forward-backward) were inextricably linked to time. Masses (or concentrations of energy, which he showed to be equivalent some 10 years earlier) distort these coordinates in a particular way. In Einstein's view, embodied in his General Theory of Relativity, "gravity" was something of a phantom induced by the curvature of spacetime caused by masses pulling on the fabric of the universe itself. The theory correctly accounts for the magnitude of the apparent precession of Mercury's orbit, one powerful factor motivating acceptance of Einstein's theory.

Le Verrier and others who worked on the problem of Mercury's inconsistent orbital motion were literally chasing an answer to a problem that didn't formally exist. On the other hand, the gross irregularities of Uranus's motion didn't require knowing about general relativity in order to correctly understand them, and the prediction of Neptune's existence was perhaps the grandest Newtonian accomplishment of them all. Astronomers are to be forgiven for suggesting a new planet inside Mercury's orbit, bearing in mind their hubris in thinking that all there was to be known about gravity had been discovered by the time Le Verrier carried out his analyses of the orbits of both Mercury and Uranus.

Believing the intra-Mercurian planet to be as real as Neptune was before its discovery, Le Verrier confidently suggested it should be called "Vulcan" after the Roman god of both beneficial and destructive fire, including the fire associated with volcanoes. It was a fitting name for an object that would never wander far from the scorching heat of the Sun. By the mid-nineteenth century, the existence of Vulcan was widely recognized, and it appeared in astronomy textbooks and charts of the day (Fig. 8.3).

Le Verrier's success in deducing Neptune's orbital elements and predicting its position on the night sky—even though now understood to be a combination of skill and luck—and the apparent predictive perfection of Newtonian mechanics compelled astronomers to start a search for Vulcan in a belief that it was only a matter of time before the new planet would be found. Almost immediately, that time seemed especially short.

In late December 1859, Le Verrier received a letter from one Edmond Modeste Lescarbault (1814–1894), a physician and amateur astronomer from the village of Orgères-en-Beauce, some 70 km southwest of Paris. Lescarbault wrote Le Verrier that he had read of his work on the intra-Mercurian planet, which reminded him of something he had seen earlier in the year from his small private observatory. On March 26, Lescarbault wrote, being quite unaware of Le Verrier's prediction, he witnessed the transit of a small, dark object across

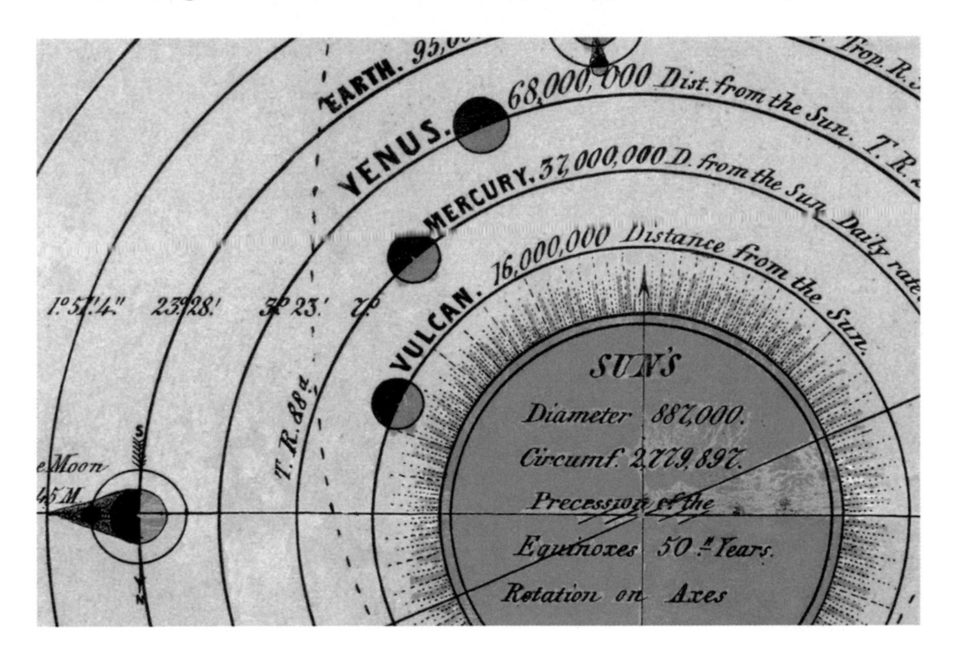

Fig. 8.3 Vulcan appears interior to the orbit of Mercury on "A plan or map of the Solar System projected for schools & academies," published at Rochester, New York, in 1846

the face of the Sun.[16] Le Verrier promptly booked a train ticket to Orgères-en-Beauce and showed up unannounced at Lescarbault's home to confront him in person for details.

No doubt surprised to see the famous Le Verrier on his doorstep, Lescarbault proceeded to describe what he saw in detail. On the aforementioned day in March, he said, he trained his 3.75-inch refractor on the Sun for some routine observations of sunspots. Among the markings on the Sun's disc that day was one that stood out. It was small and consummately round, showing none of the usual observational characteristics of sunspots. Lescarbault followed it for a while, and after a time he noticed that it was moving with an apparent speed greater than the rotation rate of the Sun. It reminded him of the Mercury transit that he watched 14 years earlier, but he knew then that Mercury was some 16° east of the Sun on that day.

He suspected it was an unknown planet orbiting somewhere interior to Earth, but the idea then seemed preposterous to him. Yet there it was, behaving for all the world just like an inferior planet during an ordinary transit. On the

[16]1879, *The Spectator* 52, 336.

outside chance it was real, Lescarbault made a measurement of the object's position and rate and direction of motion, estimating that the time required for the object to completely cross the Sun's disc was 1 hour, 17 minutes and 9 seconds.

Le Verrier found the man to be a credible eyewitness who had gone to great lengths to preserve something quantitative about what he claimed to have seen, and he left Orgères-en-Beauce satisfied that Lescarbault had, in fact, seen a bona fide transit of his predicted intra-Mercurian planet. A few days later, just after the start of the new year 1860, Le Verrier stood before a meeting of the Académie des Sciences in Paris and announced the discovery of Vulcan. From the details of Lescarbault's transit observations, Le Verrier computed an orbit for the new planet. It followed a nearly circular path around the Sun, inclined at an angle of 12° and 10 minutes of arc with respect to the ecliptic, completing one lap in 19 days and 17 hours. At a mean distance of 21 million kilometers from the Sun, it never wandered on the sky further than 8° from our star. At such a relatively small angular separation, it was little wonder that no one had seen it before.

We now know that none of this is necessarily crazy. Having discovered thousands of exoplanets beyond our Solar System in only the last quarter century, astronomers are unfazed by the notion that a planet would orbit its host star so closely that it could complete one circuit in only a matter of days. The space between Mercury and the Sun is remarkably large and compares favorably to the length scales between the giant planets and the orbits of their larger moons. The American astronomer Daniel Kirkwood (1814–1895) considered the problem in 1847 and found no trouble inserting an otherwise unseen planet into the gap:

> The distance from the centre of Jupiter to the nearest satellite is about three times the equatorial diameter of the primary. If, therefore, we suppose the distance of the nearest primary planet to have the same ratio to the diameter of the sun, the orbit of such planet will be somewhat less than 3,000,000 miles from the sun's centre. Consequently, in the interval of 37,000,000 miles there may be four planets, the orbit of the nearest having the dimensions above stated, and their respective distances increasing in the ratio of Mercury's distance to that of Venus.[17]

It might be a stretch to extrapolate the existence of a previously unknown object so close to the Sun to be forever lost in its glare—except those instances

[17] *Literary Record and Journal of the Linnæan Association of Pennsylvania College*, 3, 131.

when it passed directly in front of the Sun as seen from Earth—but the reasoning that anticipated its existence was based on well-understood physics. If science could correctly predict the existence of one new Solar System planet, why not another? In fact, if we accept the validity of the Newtonian theory, and knowing nothing of the physics that lay beyond it, logic seemed to demand that the burden of proof should be reversed: given the observed precession of the perihelion of Mercury's orbit, until someone could show that *no* such interior planet existed, we should proceed from the assumption that it is real. Le Verrier's startling announcement brought some minor fame to Lescarbault, who was made a Chevalier of the Légion d'Honneur for this services to the Republic, and invited to speak at a number of French learned societies.

Le Verrier had his detractors, some of whom openly discounted the reliability of Lescarbault's "discovery." One such figure, Emmanuel Liais (1826–1900), claimed to have observed the Sun at the exact same moment as Lescarbault while in the employ of the Brazilian government as an astronomer in Rio de Janeiro. Despite using a telescope twice as powerful as Lescarbault's small refractor, Liais reported noticing nothing vaguely like a transiting planet. Representing himself "in a condition to deny, in the most positive manner, the passage of a planet over the sun at the time indicated,"[18] Liais insisted that he saw nothing of the sort. But, as Kirkwood wrote in his account for *Popular Science* in the autumn of 1878,

> the astronomer of Brazil did not stop with denying the truth of Lescarbault's observations. He boldly called in question the conclusion derived by Leverrier himself from a laborious discussion of the observed transits of Mercury. ... This is nothing less than the charge, by an eminent astronomer, that the observations and measurements claimed by Dr. Lescarbault were a pure fabrication.[19]

Le Verrier's celebrity status as the discoverer of Neptune may have unduly influenced the amateur astronomers who began to write to him with their own accounts of otherwise unexplained transits of what they were sure must have been Vulcan. Kirkwood continues:

> The sun was again watched during the last days of March in 1860 and 1861, in the hope of reobserving the new member of the system. The search, however, was unsuccessful until March 20, 1862, when Mr. Lummis, of Manchester, England, between eight and nine o'clock a.m., observed a perfectly round spot

[18] 1878, *Popular Science*, 13, 733.
[19] *Ibid.*

> moving across the sun. Having satisfied himself of the spot's rapid motion, he called a friend, who also noticed its planetary appearance. From these imperfect observations two French astronomers, MM. Valz and Radau, computed elements of the planet: the former assigning it a period of 17 days 13 hours; the latter, one of 19 days 22 hours.[20]

Then the magic seemed to suddenly end: "From 1862 to 1878 the planet was not seen, or at least no observation was well authenticated."[21] Nor were any observations reported at all in 1861. The claims made in 1860 and 1862 were later shown to be entirely unreliable; some were reports of sightings made years earlier on uncertain dates, which in any case yielded no useful information to help refine Vulcan's orbit.

Le Verrier extracted what he could from the mess of data and predicted dates of future Vulcan transits of the Sun. But the amateur astronomers who waited patiently with their telescopes pointed at the Sun on the appointed days and hours were crestfallen when each and every opportunity turned up precisely nothing. Prospects for Vulcan's objective existence seemed to fade. While the idea was not abandoned within the scientific community, steadily fewer astronomers bothered to look for it. Nevertheless, Le Verrier went to his grave in 1877 convinced that he had correctly predicted the existence of not one but two new planets of the Solar System.

On July 29, 1878, the path of a total solar eclipse cut a wide swath from northwest to southeast across North America from Alaska to the Caribbean. Two American astronomers, James Craig Watson (1838–1880), the Director of the Ann Arbor Observatory at the University of Michigan, and Lewis Swift (1820–1913), an amateur from Rochester, New York, went "out West" to see the first total eclipse visible from the United States in a decade. Both men mounted their respective eclipse expeditions for the express purpose of searching for Vulcan; the eclipse itself was just a means to an end that temporarily attenuated the intense glare of the Sun.

Being predisposed to the notion of Vulcan's reality, it comes as little surprise that both reported seeing something close to the Sun that they thought might be it: a dim "star" that appeared out of place in the darkened sky during the few minutes of totality. Watson, observing from Wyoming Territory, described an object with a visual magnitude of +4.5 some two and a half degrees southwest

---

[20] *Ibid.*

[21] One possible exception is the report of Aristide Coumbary (?–1896), a French astronomer at Constantinople, who reported the transit of a small planet on May 8, 1865. See Coumbary and G. F. Chambers, 1865, Observation of a Supposed New Inferior Planet, *Astronomical Register*, 3, 214.

of the Sun. Swift, at an observing station to the south in Denver, Colorado, guessed the angular separation between Sun and object was about 3°. He used the fifth-magnitude star θ Cancri, a few minutes of arc away, as a reference, describing the interloper as about the same brightness and lying very nearly along a line joining the star and the center of the solar disc.

Watson and Swift were both highly reputable observers, unlikely to have mistaken a background star for a planet during the darkness of a total solar eclipse; in fact, by 1878 Watson had already visually discovered 22 asteroids, and the eagle-eyed Swift had found several comets that bore his name. Their reports reinvigorated interest in Vulcan: if not the hypothesized planet, then what *did* they see? And why had no one noticed it during previous total solar eclipses? Both Watson and Swift described the color of the object they saw as distinctly red, and Watson even stated that it showed a discernible disc in the telescope eyepiece. That strongly suggests a planet, rather than a star, as the latter are much too distant to appear resolved in small telescope views. But the small disc suggested nothing like a strong phase that one would expect if the object were in the part of its orbit nearer to the Earth, and from this Watson inferred that it must be approaching superior conjunction on the far side of the Sun at the time of the eclipse.

But the little red visitor was not the only oddball that the two observers noted in 1878. Each saw another object in a different part of the sky that he believed did not match known stars. Analyzing his observations after the eclipse, Swift found an error in the coordinates for both objects he saw, leaving the whole state of affairs in an even more confused condition because now the two observers claimed *four* anomalous objects between them during one eclipse.

The scholarly world greeted the claims with skepticism, while Watson's chief scientific rival, the German-American astronomer Christian Heinrich Friedrich Peters (1838–1890), used it to make Watson an object of ridicule. Although he was already disinclined to believe in the reality of Vulcan, Peters thought he could prove that every otherwise unexplained object seen by Swift and Watson was a thoroughly ordinary background star. Peters contended that in his haste to record data during the brief 2 minutes and 40 seconds of totality, Watson simply erred in reading the position circles on his telescope. His treatment of Swift was harsher, as he dismissed the entirety of his claims with the implication that he made the whole story up after reading a published account of Watson's observations. Swift, however, leaned on his own sterling reputation and got the last word: "Astronomers are left no alternative but to conclude that I saw at least one, and probably two, intra-mercurial planets."[22] Ultimately, history disagreed.

Astronomers continued looking for Vulcan during total solar eclipses as late as 1908, but no candidate yielded any repeat observations.[23] Ironically, it during was the total solar eclipse in 1919 that the first direct evidence for Einstein's theory of general relativity emerged, as astronomers measured the deflection of light from background stars as it passed through the edge of the Sun's "gravity well" on May 29. Rays of starlight, constrained to travel along warped spacetime near the Sun, bent by exactly the amount that Einstein's theory predicted to within experimental uncertainties, confirming the underlying physics. This development, piled atop the successful calculations for the precession of Mercury's orbit, drove the last stake through the heart of Vulcan.[24]

The moral of the story of Vulcan, like that of the Martian canals (Chap. 1), is that smart people can convince themselves of the reality of something that is no more than a mere suggestion of possibility, even if that possibility is backed up by sound scientific reasoning. That something might exist doesn't make it so, no matter what people think they see. And, like the Martian canals, Vulcan really *was* backed up by more than plausibility—until Einstein showed Vulcan was never required in the first place. For the canals of Mars, it was the eminence of an observer who stood by what he claimed to have seen, while in the case of Vulcan, belief was bolstered by the successful prediction of the existence of Neptune before its discovery. Both claims relied on inadequate skepticism in the face of scientific celebrity.

Sometimes faith in the authorities of science is vested in their reputations as individuals, while in other cases the devotion stems from adherence to

---

[22] Eggen, O. 1953. *Astronomical Society of the Pacific Leaflets*, 6(287), 291.

[23] Yet some people refused to give up, convinced of the reliability of eyewitness reports of "Vulcan" transits in the historical record. Among these was Henry C. Courten, of Dowling College, New York, who studied photographic plates made of the sky near the Sun during the total solar eclipse of March 7, 1970. Courten claimed the detection of several objects on the plates that seemed to be on small, Sun-centered orbits. After dismissing a few of the candidates as photographic artifacts, Courten was left with the conclusion that at least seven of the objects were real. He came to believe in the existence of at least one intra-Mercurial planet with a diameter between 130 and 180 km, orbiting the Sun at a mean distance of some 15 million kilometers. The other objects were thought to be constituents of an asteroid belt interior to Mercury. None of Courten's claims were subsequently verified. For a detailed account, see Seargent, D., 2011, *Weird Astronomy: Tales of Unusual, Bizarre, and Other Hard to Explain Observations*, New York: Springer, 68.

[24] While it's possible, and even likely, that very small asteroids exist interior to Mercury's orbit ('Vulcanoids'), they must be so small as to have avoided detection to date. Searches have ruled out any such asteroids larger than about 6 km in size. See, e.g., Steffl et al., 2013, A Search for Vulcanoids with the STEREO Heliospheric Imager, *Icarus*, 233(1), 48–56; and Merline et al., 2016, Search for Vulcanoids and Mercury Satellites from MESSENGER, 47th Lunar and Planetary Science Conference, held March 21–25, 2016 at The Woodlands, Texas, LPI Contribution No. 1903, p. 2765.

incomplete ideas that are ultimately inadequate descriptions of the physical universe. The Ashen Light may sample a little from both sources. To test that idea, we first have to ask whether our eyes simply deceive even the best observers among us.

9

Perception Revisited: The Psychophysics of the Ashen Light

Although long thought perhaps the simplest form of mental representation, belief turns out to be a powerful filter through which the human mind interprets information about the world received through the senses, influenced as much by social reinforcement as our own firsthand experiences. Whether the conviction of the reality of the unseen in religious faith or the ability of the brain to form reliable, predictive "models" of the world based on past observations, belief forms a cornerstone of the concept of conscious thought. In short, belief exists at the confluence of the dual streams of emotion and cognition, a mixture of responses from the nervous system and the thinking brain.

We have seen how even eminent scientists can hold fundamentally erroneous convictions about the nature of the universe simply because those convictions appeal to what they already believe to be true. The quest for objectivity demands that we decouple observation and data interpretation from intuitive notions about how nature is or ought to be—or at least agree to correct our beliefs in that regard when they are contradicted by experience. But before new information can modify beliefs, an intermediate step takes place: perception.

The nervous system accepts inputs from the sensory system, augmented by interpretation in the brain based on expectation, memory, and adaptation through learning. Certainly many phenomena that produce sensory signals are entirely independent of human observers, notwithstanding whether a tree makes a sound if it falls in a forest and no one is around to hear it. Most psychologists and philosophers accept the independent existence of the stimuli

J. C. Barentine, *Mystery of the Ashen Light of Venus*, Astronomers' Universe,
https://doi.org/10.1007/978-3-030-72715-4_9

that humans receive, process, and interpret every day. The branch of science called psychophysics aims to quantitatively describe the relationship between the physical nature of sensory input and how it maps to perception. And no thorough examination of the Ashen Light—a distinctly visual phenomenon—is complete without an excursion into that subject.

The Effects of Observer Bias

What if the Ashen Light has no objective basis in reality and is simply the result of biased observers primed to see what they already believe to be true? In that case, it wouldn't be a matter of perception leading observers astray, but rather the other way around: those observers' own predilections shape sensation in something of a feedback cycle between brain and eye. There is some evidence for this in psychology. For example, people who believe in paranormal phenomena tend to report experiencing them more than those who don't.[1] A comparable effect is found among people reporting ecstatic or mystical religious experiences: they're overwhelmingly believers. These are examples of *confirmation bias*, which results from the influence of desire on beliefs.

"To whatever we might attribute" the Ashen Light, the American astronomer Simon Newcomb wrote in 1902,

> it ought to be seen far better after the end of twilight in the evening than during the daytime. The fact that it is not seen then seems to be conclusive against its reality. The appearance illustrates a well-known psychological law, that the imagination is apt to put in what it is accustomed to see, even when the object is not there.

Newcomb even suggested that the familiarity of the Earthshine, and the similarity in the appearance of the lunar and Venus crescents, led directly to the Ashen Light: "We are so accustomed to the appearance on the moon that when we look at Venus the similarity of the general phenomena leads us to make this supposed familiar addition to it." James Muirden wrote that the

[1]There is an extensive literature on the subject. For more recent examples, see, e.g., Wesiman et al., 2002, An investigation into the alleged haunting of Hampton Court Palace: Psychological variables and magnetic fields, *Journal of Parapsychology*, 66(4), 387–408; M. van Elk, 2013, Paranormal believers are more prone to illusory agency detection than skeptics, *Consciousness and Cognition*, 22(3), 1041–1046; and Dagnall et al., 2016, Paranormal Experience, Belief in the Paranormal and Anomalous Beliefs, *Paranthropology*, 7(1), 4–15.

propensity to see light on the unlit side of Venus is "treacherously easy, … for some ocular effect makes the area between the horns appear lighter in tone than the outside sky, and a careless observer might well record this as a 'sighting'." Thomas Dobbins labeled the tendency of the eye and brain to complete the figure outlined by the Venus crescent "insidious."

Perceptual psychology supports this notion. Organisms are constantly bombarded by sensory stimuli in their environments. In complex animals, among which decision-making is a critical survival skill, optimizing this process involves handling a lot of information from multiple senses over varying lengths of time. Past experience informs the reliability of that information, helping to cut down how long it takes to decide how to react.

Think about crossing a busy street with a lot of vehicular traffic. On clear days, we look for traffic more than we listen, since we can generally see cars coming long before they approach us. But what about on a foggy day, when visibility is low? As visual information becomes less useful, we might well switch over to listening for the sound of cars. It puts us at a disadvantage, given the limited distance over which the noise of oncoming traffic is likely to be heard, but some warning is obviously better than none. People who lose a sense, such as those who go blind or deaf during their lives, often become hypersensitive to their remaining senses as the brain's circuitry rewires itself around the parts that no longer interpret sensory input.

"Humans often evaluate sensory signals according to their reliability for optimal decision-making," wrote Benedikt Ehinger, a postdoctoral researcher at the Donders Institute for Brain, Cognition and Behaviour in the Netherlands, and coworkers in 2017.[2] But when people process incomplete information, they often reach the wrong conclusion first: "a percept that is partially inferred is paradoxically considered more reliable than a percept based on external input." In other words, people have a tendency to "fill in" missing information under such circumstances, and their intuition treats those half-truths as more real than authentic experiences.

The same effect readily extends to vision in the most familiar way. Signals from the human retina are conducted to the brain through the optic nerve, which attaches directly to the retina at the back of the eyeball. At the point of attachment, there are very few photoreceptors, yielding a "blind spot" just slightly off of the center of the visual field in each eye. The brain simply receives no significant stimuli from this part of the field, but we don't perceive any discontinuity there when we view the world. In fact, it would be a significant

[2] *eLife*, 6, e21761.

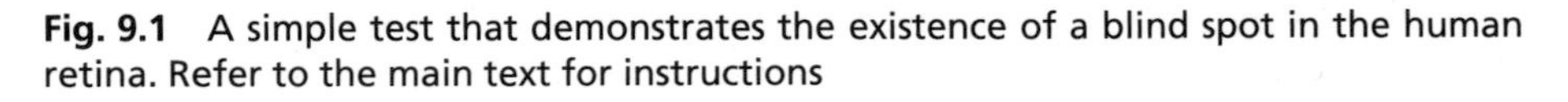

Fig. 9.1 A simple test that demonstrates the existence of a blind spot in the human retina. Refer to the main text for instructions

disadvantage to our survival if some part of our view of the world were permanently obscured. So, over billions of years, evolution figured out a clever trick: the brain interpolates information from photoreceptors bordering the blind spot and presents a seamless image.

Only through deliberately tricking the brain into betraying the blind spot's presence can most people notice it; Fig. 9.1 shows a way of revealing it. Hold this book at typical reading distance with your nose centered on the space between the cross and circle. Cover or close your left eye and stare at the cross with your right eye, and then slowly draw the book toward your face. Keep your right eye on the cross. When the page is somewhere around 25 to 35 centimeters from your face, the circle will disappear from view. But look carefully—in its place is the white of the surrounding page. Your brain is making a guess at what might be in the place occupied by the circle, and it guesses "white."

The same works for color vision. Look at Fig. 9.2, which repeats the first test but adds both color and spatial patterning. Repeat the same procedure, focusing on the cross with your right eye while covering or closing your left. At just the right distance, not only will the yellow circle disappear to the right eye, but in its place is… a red circle. If the brain only looked at the immediate region of the yellow circle –its "surround"– it should fill in white. Instead, it reaches a little further, senses the grid of red circles, and presumes that the missing information should follow the suggested pattern. In natural settings, the brain is often correct. It knows where to look and finds what it expects to see. With more practice, the trick becomes easier to repeat as the brain becomes conditioned to the array of red circles.

Writing in 1971, William Napier identified both processes as potentially at work in the case of visual observations of the Ashen Light. "First, one knows where to look for the phenomenon," he wrote. And, "second, it is well known that visual acuity increases with practice at the eyepiece of a telescope: the Ashen Light is usually looked for by experienced amateurs."

Look for it they did, but some refused to believe their eyes. Henry McEwen, Director of the British Astronomical Mercury and Venus Section for a remark-

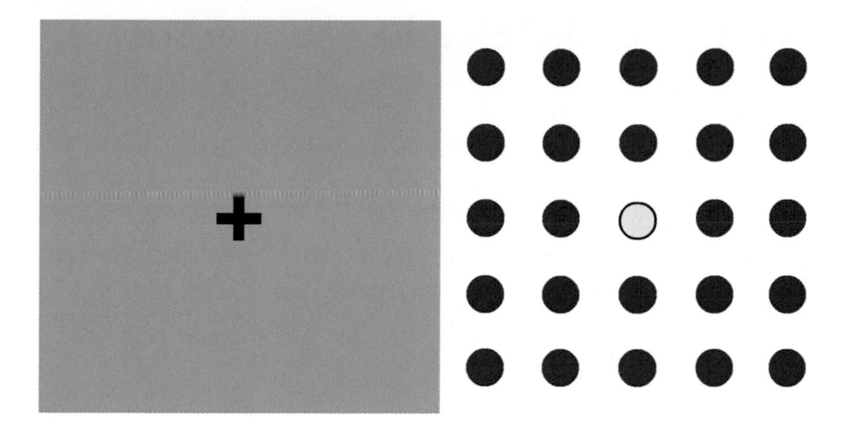

Fig. 9.2 A color version of the same test as in Fig. 9.1

able six decades, quoted a 1927 observation of the Ashen Light by one C. S. Saxton of Somerset, of whom little is known. As previously related in Chap. 6, Saxton reported that "when scrutinizing the horns in an attempt to determine their full extent as exactly as possible" on September 10, the day of inferior conjunction, "the observer was very much astonished at being able to trace faintly the outline of the entire circular disc against the dark blue background." He noted his reluctance to expand on his initial report as the result:

> of a very strong scepticism with regard to previous observations of the same kind, the explanation of which has always seemed easier by assuming some kind of optical illusion, rather than by regarding them as objectively real. The writer is now in the disagreeable position of having no satisfactory explanation at all to offer, either on the basis of illusion or of actual objective reality, and therefore gives the following account of the observation without submitting any theory to explain it, but rather in the hope that some other Member of the [British Astronomical] Association may be able to put forward an explanation which will obtain more or less general assent.[3]

What drives people to explicitly *distrust* the information their senses present to them or, as in this case, to suspect imperfections in the instruments used to amplify sensory stimuli? And where does the apparent disconnection between physical stimuli and perception occur that can lead us to the wrong conclusion?

---

[3]1927, *Journal of the British Astronomical Association*, 37, 345.

Down the (Duck-)Rabbit Hole of Vision Science

For answers to these questions, I contacted Manuel Spitschan. At the time we talked, Spitschan was a postdoctoral researcher in Psychiatry and Behavioral Sciences at Stanford University; since then, he's moved on to a Sir Henry Wellcome fellowship at Oxford. He's an expert in how non-image-forming visual systems work, interested in how light affects the brain and behavior beyond what we consciously see and perceive. When I reached him by Skype at his office at Stanford, I was ready with a laundry list of questions about the limits of our current understanding of human vision. What followed was a freewheeling and at times deep discussion with surprising relevance to the Ashen Light.

I prepared Spitschan in advance of the call with an email on the background of the Ashen Light: what we think we know, what we're sure we don't know, and which elements suggested some intersections with vision science. After exchanging pleasantries, we got right down to business, and he dropped his first metaphorical bomb in terms of how to interpret what eyewitnesses of the Ashen Light were telling us: perception is "immensely variable." And that remains the case even when you fully understand the physical nature of the stimulus, and you consider higher levels of perception. In other words, he knew right from the start that he had no definitive answer to my underlying question.

As a vision scientist, Spitschan told me that he found the Ashen Light interesting as a phenomenon in which it's entirely possible that the brain is filling in information where it doesn't actually exist. "It takes a huge amount of neural information processing to construct a robust and consistent perception of the world," he said. In addition, no stimulus exists in perfect isolation. Think of the different colors on a computer monitor, for example, which can look very different to the eye depending on what fills the visual field beyond it. (A white wall? A shelf full of books? A window looking outside on a sunny day?)

"There are a whole lot of contextual effects going on there. And sometimes, the mind's expectation of seeing something in particular modulates the percept" he said, tossing out a psychophysical term for the subjective interpretation of a physical stimulus. "Ambiguous" images are good for showing this effect. I asked for an example.

He paused. "Have you seen the duck-rabbit before?"

I drew a blank. "I'm going to go with… 'no'?"

"Look it up on the Internet while we're chatting," he suggested.

Fig. 9.3 The "duck-rabbit," a classic example of a bistable or ambiguous image

An example of a bistable figure,[4] the duck-rabbit first appeared, unattributed, in an 1892 issue of a German humor magazine called *Fliegende Blätter* ("Flying Leaves"). The original engraving (Fig. 9.3) at first looks like the disembodied head of a duck, facing left and tilted upward, its beak parted slightly as if in mid-quack. But look at it again and see instead the head of a rabbit facing right, the duck's beak becoming its long ears. "Welche Thiere gleichen einander am meisten?" ('Which animals are most like each other?') its caption asked. Beneath the drawing, the continuation answered: "Kaninchen und Ente" ("Rabbit and Duck"). In the picture, they're the same thing at the same time. Which animal one sees depends on what one looks for. Looking for a rabbit? You'll see it. Looking for a duck? It's there, too.

Another way of thinking about it, familiar to readers of popular science, is the thought experiment known as "Schrödinger's Cat." It was proposed by the Austrian physicist Erwin Schrödinger (1887–1961) in 1935 to draw attention to what he believed to be an important shortcoming of the dominant interpretation of quantum physics when applied to macroscopic systems. In the experiment, a cat is placed in a sealed box along with a glass container of poison, a piece of radioactive material, a radiation detector, and a hammer. Since radioactive decay is a random process, at any moment after the box is sealed, an atom of the material may decay that trips the detector. In turn, the

[4]Spitschan was quick to point out that bistable figures are *not* optical illusions, because they really do show both aspects claimed. For more on this phenomenon, see, e.g., Long, G., and Toppino, T., 2004, *Psychological Bulletin*, 130, 748–768.

detector actuates the hammer, shattering the glass vessel, releasing the poison, and killing the cat. The dominant "Copenhagen" interpretation of quantum physics insists that, some time after the box is sealed, the cat is literally equal parts alive and dead; it is said to be in a "superposition" of the two states. Furthermore, it remains in this superposition until the box is opened, and one or the other possible ending condition is observed. At that point, the superposition is said to have "collapsed" into a single reality.

Clearly no one believes in a simultaneously half-dead and half-alive cat, because it violates every experiential notion we have about the macroscopic world. But the experiment has shown that systems on the scale at which quantum mechanics takes over from the familiar realm of "classical" physics really *do* behave this way. In the weird and often non-intuitive quantum world, mathematics implies that systems can be found in situations where their various possible states are tangled up together. It is only when the observer interrogates the system by interacting with it somehow—opening the box and looking inside, as it were—that quantum superposition ends and the system collapses into one, and only one, final state.

Although this is a naïve and simple exposition on Schrödinger's Cat, it has real relevance to our brains' experience when presented with visual stimuli like the duck-rabbit. Faced with ambiguous visual input that involves familiar elements, the brain quickly collapses the image into only one of those elements and discards the alternate possibilities. How it does this, and why it often happens instantaneously, is still not well understood, but it seems to follow a trajectory outlined by the brain's predisposition to see one element preferentially over the others. What determines this?

The way to get at the nature of a percept, Spitschan suggested, is to carry out experiments where subjects aren't necessarily primed to see anything, and ask what they see. If the subjects aren't told what they're supposed to see, and instead are shown random visual "noise," they will generally report sensing something with a defined structure. The human brain is just really good at picking out patterns, even when there's nothing actually there. When confronted with something like a bistable image, the brain forces itself to choose from competing versions of reality.

"Okay, then," I said, "what about the faint end of things? What happens when the eye is working near the limit of vision?"

"The eye is really underrated as a detector of faint light," Spitschan answered. "It's impressive that the retina, as this biological thing, can detect single photons. But most people never become that completely dark-adapted." Furthermore, with sensitivity to faint light comes poor spatial resolution due to the extent to which the highly sensitive rod cells are distributed across the retina. In particular, they're least dense in the fovea, the region in which we have the best sense of spatial discrimination. It's also where almost all of the cone cells concentrate, meaning that our sense of color perception drops off rapidly in the periphery of our sight.

I stopped him. "Wait, what's the evolutionary advantage of that? So what, we can see faint light kind of okay, but then we can't identify what we're seeing?"

"Bear in mind that there just aren't a lot of animals that have only cones. Either it's rod only or rod/cone." It's most advantageous to have rod cells in your retina when the brightness of objects observed is between about 0.001 and 3 candelas per square meter, which turns out to be rather close to a lot of the nighttime situations we encounter under artificial outdoor lighting. Although cones give us good spatial resolution and color information about whatever it is we're seeing, they suffer from a lot of noise in the form of spurious signals at low intensity that form something of an indistinct background level. Once the ambient light level drops to the lower end of the photopic range of intensities, near one candela per square meter, one quickly hits a sensory "floor" established by the noise in individual cone cells.

Think about being outdoors about an hour after sunset. That's when typical ambient light levels are sufficiently low that your cones are basically giving up and your rods are starting to take over. Until the night becomes fully, astronomically dark, and even then in the absence of brighter sources of light, whether natural or artificial, the rods and cones team up to offer a compromise: mesopic vision, which involves some ability to tell colors apart (but not much), limited spatial resolution, and better sensitivity to faint things—especially those that are moving slowly through the field of view. I remembered the latter from my own experience looking at faint objects through the telescope eyepiece, having long since noticed that rocking the scope back and forth a little helped lift the faintest galaxies out of the light of the background sky just long enough to convince my brain that I actually saw them.

"So the evolutionary advantage to both mesopic and scotopic vision is maintaining a large dynamic range of response to stimuli down to fairly low illuminances," he said. "But then again, I don't think rod vision is all that great." Well, no, I thought, but it might have helped our distant ancestors stay alive.

Rod vision is distinctly blue-sensitive. "Why is that?" I wondered aloud.

"That's a great question," Spitschan said. There are basically two lines of reasoning resulting from experiments. Rod cells get their sensitivity to light from a molecule called rhodopsin, which is occasionally referred to by the much more poetic name "visual purple." Rhodopsin is exquisitely sensitive to light, initiating signals from rods to the brain even at incredibly low illumination levels. He noted that the same molecule is present in the retinas of certain fish species. In the 1970s, scientists pulled fish from various depths of Lake Baikal, the world's deepest freshwater body, and looked at the spectral sensitivity of the rhodopsin in their eyes as a function of the lake depths in which they tended to dwell. It turns out that the sensitivity of the fish rhodopsin varied according to the depth from which the fish were taken.

The British zoologist John Lythgoe recounted the story in his 1979 book *The Ecology of Vision* (which "has a Victorian feel to it," Spitschan added), arguing that there are intrinsic matches between an organism's phenotype, or the characteristics of individual organisms resulting from the interaction of their genetic inheritance with their natural environments, and those same environments. Among the strongest themes in Lythgoe's book is that the spectral sensitivities of the photoreceptors in individual species are no accident, but rather have been guided through evolution according to the nature of the visual spaces they typically inhabit. Evolutionary pressures drive them toward or away from sensitivity to specific wavelength ranges. Humans are no exception to this rule.

We see evidence for that shift as our visual systems transition from cone-dominated to rod-dominated under decreasing ambient light levels. This is most familiar in the way colors around us seem to change during evening twilight. As night falls, cones respond less to visual stimuli and rods start to take over. But a curious thing happens as the changeover from photopic to scotopic vision occurs, and it's due to the difference in spectral sensitivity in the two types of photoreceptors. While the rods render essentially "black and white," or grayscale, images because of their inability to distinguish colors, their intrinsic blue sensitivity means blue things have a higher perceived brightness than, say, red things.

The result is that colors in the photic environment seem to disappear at different rates depending on the wavelengths of light they represent. Reds fade from view first, followed by greens. Blue tones are the last to vanish from human visual sensation at low light levels. That means, for example, some flowers actually look brighter in twilight than they do under full sunlight. It also explains why moonlight looks "cool" or bluish to the eye, whereas it's actually redder than sunlight.[5]

Color perception itself, therefore, is neither rabbit nor duck, but a different animal entirely. "In the last 15 years," Spitschan said, "adaptive optics has

[5]For a detailed discussion, see Ciocca, M., and Wang, J., 2013, *Physics Education*, 48(3), 360–367.

come to vision science. We're now stimulating individual cones and getting information about their sensitivity to light intensity and color. But things get complicated there." It happens to be the case that our sense of color perception depends on more than just the kind of cone on which an individual photon happens to land, but also the types of cones surrounding it. The sensation of color doesn't have a simple, one-to-one relationship to the incoming stimulus.

"What about the reports by some observers that the Ashen Light seemed to take on a 'coppery' or reddish color, especially against daytime sky?" I asked.

"It's really difficult to go from wavelength to perceived color, to predict what the seen color is," he explained. "There are models out there that try to get to this, that try to predict the seen color given a set of cone excitations."

I pressed him on that point. "So it doesn't tell you much about the actual source of the light?" I thought back to the example related in Chap. 5 in which two otherwise invisible near-infrared photons landing on the same retinal photoreceptor at the same time produce the sensation of "green" in the observer's brain.

"No," he said. "Processing of color happens by antagonism of a set of specific signals, and a post-receptoral channel is doing the signal processing. We call it 'spectral opponency' in the primary retina." It evolved as early as some fish species and is the machinery for color discrimination. The brain looks at differences between the signals coming from certain cone types, labeled according to the wavelengths of light to which they're most sensitive: red-sensitive ("long wavelength") cones are called "L", blue-sensitive ("short wavelength") are "S", and ones mostly sensitive to green light are called "M" ("medium wavelength"). The $L - M$ channel, for example, mediates our sensitivity to differences along a red-green axis. There's also an $S - (M + L)$ channel that can lead to a bluish percept; a decrease in $S - (M + L)$ is seen as yellow.

"That happens in the retina? Or in the brain?" I asked.

"The retina itself," Spitschan said, "or, at least, as early as the retina." By that he meant that some signal processing takes place *in* our eyes, before the signals ever reach the brain. "The processing of color happens at multiple levels."

As an example, he described cases reported in the scientific literature involving people who sustained damage to parts of their brains used in the "backend work" of color perception. Sure enough, the people lost their ability to sense color altogether. In other situations, the neural wiring of the brain is intact, but people suffer degeneration of their retinas and lose the ability to discriminate colors. Along the route from eye to brain, there is no clear point at which we can say, with certainty, that distinct colors are firmly "perceived."

That was a startling contention. If true, it implies that the human sensory system that gives us information about a fundamental aspect of the photic world around us isn't much like an electronic light sensor. Consider a digital camera, be it of the expensive, digital single-lens reflex (DSLR) variety or the type that people all over the world carry around integrated into their mobile phones. At the heart of each is a thin slab of semiconductor material that is "doped" with chemical impurities meant to render it sensitive to light in a particular way.

Generally speaking, all such detectors operate on the same principle: a photon of light lands on one of the pixels, or discrete, light-sensitive units of the detector, and generates a minuscule electric charge. Supporting electronics amplify the signal until it is measurable by an analog-to-digital converter, a device that reads an analog current or voltage and assigns it a "digital number" whose magnitude is proportional to the input signal. In its simplest configuration, such devices are photon counters, reporting one digital number unit for each photon reaching the pixel during the time in which the sensor was actively receiving a signal.

Up to this point, color hasn't entered the picture. Depending on the exact composition of the light-sensitive substrate, it may be panchromatic—essentially sensitive to all colors of light equally—or it may only "see" a limited range of light wavelengths. This has to do with the energies of the individual photons, which vary in a predictable way according to their wavelengths. It's a view into the often-weird quantum world, which considers the photon a "quantum" of energy whose observed behavior depends on how one makes the observation—like a tiny Schrödinger's Cat. Before a full understanding of all of this emerged in the early twentieth century, physicists noted that shining light onto a polished metal surface caused it to kick out electrons, but only if the light was of a certain color. Below some threshold corresponding to a particular energy of light, no electrons jumped off the metal. Furthermore, the intensity of the light didn't matter: crank up the brightness of light below the energy cutoff, and nothing happens. But shine very dim light with the requisite energy on the metal and watch the electrons spring forth.

In the world of classical physics, where light is thought of solely as a continuous wave propagating through space, this makes no sense. Albert Einstein correctly explained what was going on in a 1905 paper; this "photoelectric effect," he argued, required thinking of light as a stream of discrete units of energy. To correctly reproduce the other, wave-like features of light, it's convenient to think of photons as "wave packets," little wave trains that arrive, pass by, and move on at a particular speed.

Einstein's explanation, which won him the 1921 Nobel Prize for Physics, postulated that the electrons in metals are stuck to those materials with a certain strength that depends on the type of metal. Called the "work function" of that metal, it is the energy required to unbind an electron from the bulk metal and send it flying on its merry way. Since photons carry only a specific amount of energy that depends on their wavelength, the energy of motion of the departing electron is simply the photon energy minus the work function. Any photon energy less than the work function imparts too little *oomph* to overcome the electrostatic forces binding the electron to the metal, and nothing happens. That remains true even if one bombards the metal with light at that too-low energy. The electron stays firmly put.

Electronic sensors of light make use of this principle: too little photon energy won't dislodge electrons in the semiconductor material, and no current will flow. But any photon energy over the material's work function gets them moving and ultimately counted by the analog-to-digital converter. Changing the composition of the sensor material can therefore change the degree to which pixels are sensitive to different energies of light and hence their colors. But the digital numbers returned by individual pixels don't tell the whole story about color, because they only yield information about the flux of photons in specific colors to which those pixels are sensitive. So how do digital sensors make color images?

The answer is that they mimic the human visual system in a particular way. The cones in our retinas respond to photons across a broad range of wavelengths, but the way in which the sensitivities of the respective cone types combine results in much stronger responses to certain colors than others under bright-light conditions. Still, dial in numbers for the L (red), M (green) and S (blue) cones, and the eye and brain deliver the sensation of nearly 10 million distinct hues to the viewer. Digital sensors mimic this process by adapting the response from clusters of pixels also broadly sensitive to red, green, and blue light and returning a composite value for each cluster that includes not only the color information but the overall brightness recorded by the pixel. Reconstruction of the image on a computer monitor involves playing back this information and specifying, for each luminous element on the screen, a color and a brightness.

This isn't really how the analog eye/brain system works, however. To the point where light falls on cones, it resembles a photoelectric process mediated by light-sensitive proteins called opsins. When light strikes an opsin molecule in a cone, instead of ejecting an electron, it changes its structure. This redistributes the charge within a molecule, leading to it becoming "polarized," with the positive charges ending up mostly on one side of the molecule and the negative charges mostly on the other. In turn—summarizing many intermediate steps—this condition causes the release of a neurotransmitter called glutamate, triggering a signal conducted down the optic nerve and into the brain for interpretation. After this release, the opsin returns to its initial "depolarized" state, awaiting the arrival of the next photon that will begin the cycle all over again.

So what happens if the retina is somehow broken, and the complete cycle described above can't take place? Spitschan told me that in such cases there's still a "forward arrow" for color perception. He pointed to "The Dress," a photograph of a Roman Originals body-contoured and lace-trimmed gown that achieved Internet infamy in early 2015. Photographed professionally under studio lighting conditions, the dress is indisputably dark blue in color and finished with black lace. But an amateur photograph of the item made under ambient lighting in a clothing shop rendered the blues of the dress in very light tones and imparted a tinge of yellow to the black parts. The photo was posted to the microblogging website Tumblr, and the photographer asked the masses: is the dress black and blue, or gold and white? Within days, the image was reproduced millions of times while dividing viewers seemingly into two equally opposed–and passionate–camps.

If the retina alone resulted in our impressions of examples like The Dress, Spitschan argued, there would have been no controversy over its color: black is (and is detected as) black, and blue is blue. End of story. "But the truth is more complicated," he said. The result depends on how much of the brain is involved and exactly how complicated the cellular machinery of the retina is. This determines whether color perception is more "front-loaded" on the retina or "back-loaded" in the brain. Human vision, therefore, really *isn't* much like an electronic light sensor. "There's a lot of signal processing involved!" he exclaimed excitedly.

I shot back: "But how can it be that the same people, experiencing the same stimulus, with exactly the same visual sensing and processing equipment in their skulls, can report fundamentally different colors for whatever thing they saw?"

Spitschan turned the question around on me. "Does it matter that there's experiential content in the first place? If you understood exactly what is happening in the brain when people are looking at a given color, could you infer that their experiences were similar?"

I was immediately drawn back to a question I've thought about again and again since I was a child: do all people see colors the way I see them? Or is my experience of, say, "red" something completely different to them? For now, there's simply no way of knowing. What we call "color" seems to exist in the eye (and brain) of the beholder.

No, I thought to myself, *you couldn't know for certain that their experiences were the same*. The only way we can communicate our experiences, to describe them to other humans, is through language. If I pick up an object I see as "red" and show it to someone else, they could have an altogether different perceptual experience of the object—but one that is consistently the same each time, like mine—and still also call it "red," because it causes the same qualitative experience each time "red" is seen. The same is true if I compare the object to something else I sense as "red," since the complementary experience would also be had by anyone who looked at both the object and the comparison example. No matter the words I choose to describe the red thing, and the fact that another person recognizes the object and uses the same word to describe it, that doesn't mean we have an absolutely identical sensory experience when we see it.

Then what should we do with the similarity in the reports of so many observers of the Ashen Light who insisted that it was one or another of a fairly limited set of colors? Essentially all of them insisted that they saw something dull and grayish, tinged with a little red, or maybe yellow, or sometimes brown. But very few report seeing blue or purple, for example. That there is not wild disagreement among observers in terms of the color of the thing they say they saw doesn't mean that there's an objective mechanism underlying the emission of light that must have the same physical qualities each time it is seen. If the psychology of belief, or delusion, or something else is at play, and at least some of the purported detections of the Ashen Light have no basis in objective reality, any sense of the color of the light might tell us something important about how signal processing in the human visual system works. Or it might tell us nothing at all.

I asked Spitschan one more question before we wrapped up our hour-long call. Although I was unsure that it had any specific application to the Ashen Light, I wondered whether there was any information in the polarization state of light that was discernible to a human observer. Polarization of light refers to the tendency of light waves—when we choose to view them as waves—to preferentially wiggle in only one particular two-dimensional plane at they travel through space. Unpolarized light, on the other hand, shows no preference for the direction in which its waves wiggle.

As esoteric as this sounds, many people have direct experience with the polarization of light because viewing the world through a filter that selects out only one polarization state has some distinct benefits. In particular, materials that act in this way make for a fantastic pair of sunglasses, reducing glare from shiny things, taking the apparent reflection of sky light off watery surfaces, and increasing the contrast between foreground and background objects. Our eyes, however, don't seem to incorporate any biological materials that limit which polarization states of light tickle the visual system. I was grasping at straws, but I figured there might be something I hadn't run across that would account for some tendency to see certain polarization states better than others. Were there?

Spitschan dashed my hopes on that point. "In humans, you can make these experiments where you find some polarization effects," he started. "I don't know that there's enough evidence to say that it guides our perception in any way. There's a lot of polarized light in the environment that we don't even realize is polarized. In the animal kingdom, polarization can be an important signal. In particular coming from the sky, and it's used as a navigation signal. If you look off-axis, does the direction of the light matter? It's different, of course. I would have to think more about it." I thanked him. He offered to send along some references by email, and we ended the call.

I sat silently in my office for several minutes afterward, trying to make sense of what I had just learned. Let's say we already had in hand unbiased and independent evidence for the Ashen Light; imagine that it had been photographed a century ago, and was as real to us as the craters of the Moon. But presume, further, that it was known to be exceptionally faint, only properly detected by the silver salts of photographic emulsions or the semiconductor chips at the heart of modern digital cameras. If everyone agreed that it was hopelessly faint as seen by the human eye, what would we make of confident reports from telescopic observers that they saw it with no ambiguity?

Accurate *and* illusory?

"Visual astronomy has many drawbacks," Tony Flanders wrote for *Sky & Telescope* magazine in 2007. Chief among these, he argued, is the authenticity of visual impressions:

> How can you be sure that another observer's report is honest and accurate? Indeed, how sure can you be of what you see yourself? It's a critical issue for anybody straining to see things on the edge of visibility. If you don't try at all,

> you're greatly handicapped. Try too hard, and you start to see things that aren't really there.[6]

Flanders used as an example something I thought I understood well: the perception of color through large telescopes when visually observing emission nebulae, like the Orion Nebula (also known as "M42"). In photographs, the nebula is alive with color in swirling pinks, yellows, reds, and blues, a riot of hues attributable to atomic physics in its rarefied gases. Brightly toned images of the nebula from famous sources such as the Hubble Space Telescope splash above the fold in major world newspapers and in glossy magazines.

But the direct, human visual experience is nothing like this. Newly minted amateur astronomers are often let down at their first views of the Orion Nebula through small telescopes, which reveal only an indistinct, grayish mist spangled with a few pinpoints of light from the young, hot stars being made there. The usual explanation for this is that small apertures don't collect nearly enough light to tickle the cone cells in the retina, so the job of detecting the nebula's light is left to the rod cells, which lack color discrimination. But the cones are sensitive to light from certain emission lines in the nebula's spectrum that trace out structures different than those outlined by its stars and dust.

Flanders continues:

> Some people (including me) have also seen pink or reddish tints when observing this nebula through big telescopes. Enough people have reported this that we can be sure it's not just imagination. And there's also no doubt that a lot of the nebula's light is indeed red.

He's right. I've had the chance to see the Orion Nebula and other nebulae visually through large amateur telescopes whose apertures range up to nearly a meter in diameter and professional telescopes up to four meters. And I came away thoroughly convinced that I had seen the rich aquamarine light of doubly ionized oxygen and the cotton-candy pink of hydrogen alpha.

"But here's the kicker," Flanders wrote.

> Laboratory experiments indicate that the red light in M42 isn't bright enough to stimulate color vision. People do indeed sometimes see red at low light levels, but it's an illusion that can be created by any color light, not just red. So it's possible that the red tints of M42 are *both* accurate *and* illusory!

---

[6]"The Reliability of Visual Observing," https://www.skyandtelescope.com/astronomy-blogs/the-reliability-of-visual-observing/ (November 30, 2007).

That parts of the nebula are red to the imaging detector, and that people viewing the Orion Nebula through telescopes sometimes report the visual sensation of "red," are facts that may have no causal relationship. Maybe, Flanders mused, the lab experiments somehow get it wrong because they don't properly duplicate the real-world circumstances in which observers say they are sensing colors. "What a mess!"

In addition to the Ashen Light, Flanders cites as an example the "notorious" transient lunar phenomena that have beguiled the Moon-observing community for at least a thousand years.[7] Beginning in the pre-telescopic times, Moon watchers have reported various glows and mists in the direction of our nearest neighbor. The introduction of the telescope gave closer views of the Moon, allowing for better pinpointing of these strange occurrences.

One famous example in modern times involves the Soviet astronomer Nikolai Kozyrev, whom we met in Chap. 6. On November 2, 1958, Kozyrev was observing the Moon in the vicinity of the crater Alphonsus when he saw what he later described as an "eruption" that lasted a half-hour.[8] The telescope he used, the 1.2-meter reflector at the Crimean Astrophysical Observatory, was equipped with a spectrograph that he used to record spectra of the glow on photographic plates. His analysis showed what Kozyrev argued were emission bands due to the carbon molecules C_2 and C_3, whose signatures are often found in the spectra of comets. During the second exposure in the series, Kozyrev later noted "a marked increase in the brightness of the central region and an unusual white colour." Shortly afterward, the light faded, and subsequent spectra appeared normal.

Very straightforward explanations have been advanced for transient lunar phenomena, including outgassing from the lunar interior possibly triggered by moonquakes; electrostatic or triboluminescent processes; and even flashes of light associated with the impacts of small asteroids. The fact that historical reports of transient lunar phenomena tend to cluster in specific geographic regions of the Moon favors a physical explanation somehow relating to geology or other influences happening near the lunar surface.

Critics of the phenomenon, however, dismiss all of it as illusions of light and distortions induced by the Earth's atmosphere. They note how many

[7] For extensive catalogs of transient lunar phenomena beginning in the sixth century CE, see Middlehurst, B., et al., 1967, Chronological Catalog of Reported Lunar Events, NASA Technical Report No. TR R-277; and Cameron, W., 2003, Analyses of Lunar Transient Phenomena (LTP) Observations from 557–1994 A.D., self-published. The latter report is available on the Web at http://users.aber.ac.uk/atc/tlp/cameron.pdf. Together, the two reports tally over 2,000 alleged sightings.

[8] For a detailed account, see Alter, D., 1959, *Publications of the Astronomical Society of the Pacific*, 71(418), 46.

of the historical accounts were made by single observers, uncorroborated by observations at other locations, and how little photographic evidence exists. Much like the Ashen Light, transient lunar phenomena could be real. Or they may simply be a very *realistic* suggestion that exists only in the mind's eye.

I'm reminded of Manuel Spitschan's question: *Does it matter that there's experiential content in the first place?* Is the reality of the percept dependent on the experience it yields? What happens when confirmation bias works the other way around, and an observer is predisposed to *not* see something? Seeing it would be an obvious problem for cognition. The Reverend Thomas William Webb described his own disbelief at first seeing the Ashen Light on the last night of January 1878: "Though a frequent observer of the planet Venus through a long series of years; I have never till yesterday evening seen the unilluminated side, which presented itself rather unexpectedly, as I had not been thinking particularly about it, and was not making it an object of special examination." It seems that Webb never again reported the phenomenon after this singular observation.

Like many authors before him, and probably many yet to come, Geoffrey Kirby dismissed the Ashen Light as a form of mass delusion in his 2018 book *Our Three Wacky Inner Planets: Imaginary, Delusionary and Inhabited. Wacky and Wonderful Misconceptions About Our Universe.* "Many astronomers believe that visual observations of the Ashen Light are yet another example of Pathological Science because these claims occur right at the limit of visibility," Kirby wrote. "Just like Blondlot with his nonexistent N-rays,[9] the Ashen Light probably exists only in the imagination of an observer straining to see what isn't there."

Or does it? The answer may in fact be that the Ashen Light both is, and is not, objectively real as a perceptual phenomenon. There certainly seem to be some instances where recorded sightings are very believable in their details, whereas in other cases more than a little imagination may have been involved. Furthermore, as we have seen there are plausible physical scenarios that can explain a faint glow emanating from the surface or atmosphere of Venus that directs a detectable flux of light toward our planet. But the question of whether or not both the perceptual and physical aspects of the problem positively coincide may never be satisfactorily answered.

[9] "N rays" were described by French physicist Prosper-René Blondlot (1849–1930) in 1903 as a new form of radiation observed in laboratory experiments. While initially confirmed by over 100 other scientists, N rays were ultimately found to be illusory. The affair is often cited as a cautionary tale about the dangers of observer bias in science.

10

Epilogue: Evanescence and Evasion

"Seeing is a process of abstraction," Richard Baum wrote in 2012.

> The eye has its own peculiarities and special properties. It does not give a literal reading of stimuli; rather, it schematizes and caricatures. Even so, the problem of perception is not wholly to be defined in terms of neurology and visual physiology. In truth, what we see is the result of several physical processes that are colored by personal experience and shaped by the manner of spirit that motivates us.

He further quoted Camille Flammarion on Venus: "We may state here, more than ever that every man sees after his own fashion, and draws after his own fashion."

Consider this book as an attempt to put the Ashen Light on trial and this chapter as the closing argument. What burden of proof ought to apply? Is it the preponderance of the eyewitness evidence, or beyond the shadow of a scientific doubt? If the latter, what kind of scientific evidence would be required to finally and definitively lead astronomers to declare "case closed"? For many, it would be a reproducible photograph of the night-side limb of Venus glowing faintly against a dark background sky, shown convincingly to be an artifact of neither scattered light in the optical system nor aggressive image processing. An equivalent spectroscopic measurement might suffice instead, particularly if it could be shown that the inferred surface brightness, if integrated across the disc, reached the threshold of human visual detection. The latter point is important if we're talking about proving the existence of the phenomenon independent of human eyes to sense it and brains to interpret it.

J. C. Barentine, *Mystery of the Ashen Light of Venus*, Astronomers' Universe,
https://doi.org/10.1007/978-3-030-72715-4_10

As we have seen, while the night side of Venus glows brightly in infrared light, the relative insensitivity of human retinal photoreceptors to infrared photons argues against this route.

In our quest to convict the Ashen Light of the crime of objective reality, underpinned by a physically plausible mechanism, what is the evidence? Much of it is circumstantial, as we wait for the prosecution to produce the proverbial "smoking gun." In Chap. 5, I enumerated a list of observational facts and asserted that the correct explanation must account for all of them. They are:

(i) **Inconsistency**: Ashen Light appearances in history have been episodic and unpredictable.
(ii) **Credibility**: Some of the great visual telescopic astronomers of history were certain they saw the Ashen Light, while others said clearly that they never saw it.
(iii) **Ubiquity**: Whether or not observers report seeing the Ashen Light doesn't seem to depend on either their locations or which telescopes they use.
(iv) **Timing, part 1**: The Ashen Light is most often reported in the 4 weeks immediately before inferior conjunction.
(v) **Timing, part 2**: Ashen Light sightings are somewhat more likely near solar maximum than solar minimum.
(vi) **Vividness**: Some eyewitnesses report a wide range of colors characterizing the Ashen Light, while others report no sensation of color at all.
(vii) **Diversity**: Discoloration of the Venus disc is sometimes reported during daytime observations, although it's unclear whether this is the same phenomenon as the Ashen Light.
(viii) **Ambiguity**: No indisputable photograph or spectrum purporting to record the Ashen Light exists.

In this book, we have examined evidence from four centuries of history; the physical environment of Venus and its atmosphere; the psychophysics of vision; and even a little perceptual psychology. I asked three questions, each in turn subordinate to the one before it: First, are historical reports of the Ashen Light believable? Second, if so, do those reports indicate a real physical mechanism of some kind, intrinsic to Venus, that explains the visual detections? And third, if so, what is that physical mechanism? Another way of framing the object of this quest is to determine whether some physical process underlies and explains the Ashen Light in such a way that its existence can be completely decoupled from eyewitness-only observations. If this is indeed the case, it would move the Ashen Light fully from a subject of astronomical folklore to one of planetary science.

There is no doubt that the Ashen Light of Venus, as a phenomenon involving thousands of observers over four centuries, exists. It's unreasonable to dismiss the whole thing as some sort of recurrent mass delusion handed down through the generations. Those who have seen it are often very insistent about what they saw, even if some denied any sense of reality to its cause. Many of the naysayers are equally militant in their rejection of the Ashen Light as anything more than a plaything of the imagination; some ascribe it to aberrations in the optics of telescopes, others to wishful thinking. A few go a step further and suggest, overtly or otherwise, deliberate deceit. But the observations remain, and there are simply too many reports from too many reliable observers to reject all of them outright.

Whether the clear plausibility of Ashen Light reports implies that it has an objective physical basis is the weak link in the chain. Absent irrefutable evidence like a photograph or some other type of empirical measurement, we simply cannot rule out the possibility that while the visual *sensation* of the Ashen Light may be perfectly real, its independence exists only in the psyche of the individual who sees it. That requires a percept (apparent light on the night side of Venus); a surround (the adjacent sky against which the brightness of the night side is compared, establishing the percept by contrast); and a system that collects and senses light, transduces the resulting signal, and evaluates the information in the signal (the telescope, human eye, and human brain).

The "front end" of that system runs from the output of the telescope through the human eye and to the retina, where photoreceptive cells turn photons into nerve impulses, while the "back end" of the system tries to figure out what the impulses mean. As we saw in the previous chapter, the back end is a savvy piece of equipment that draws on intuition and memory to fill in the missing pieces when presented with incomplete information. The ability of our brains to carry out rather complex signal processing on the fly probably helped us to adapt to an often-dangerous world and survive long enough as a species to evolve the intelligence that sets us apart from every other species on the planet. But certain challenges to this system can lead to experiences in which individuals sense very strong stimuli that simply don't exist in the wider physical world, and those who seem predisposed to sense those stimuli are most likely to report sensing them.

The Ashen Light of Venus may well fit into this space. It's not like the perception of the Moon's existence, for example. If a person never saw the Moon, the existence of the Moon could be inferred from environmental cues that don't depend on vision. With the right measurement equipment, one could actually learn a lot about the Moon having never once seen it. While there are "flat Earthers" in the world, there aren't really any "Moon denialists." Maybe whether or not there is a Moon of Earth isn't a sufficiently controversial subject to attract those with just a little too much skepticism.

But the Ashen Light isn't like the Moon in this regard. Controversy has followed it like a faithful old hound dog since Giovanni Riccioli wrote of having seen "a partial circle of light … enclosing the otherwise dark side of the disc" of Venus through his telescope in 1643. I suspect that even some of the scholarly skepticism of the Ashen Light is borne out of frustration with a topic that seems forever out of proper experimental reach but also just won't go away for good. After all, people still report sighting the Loch Ness Monster nearly 1500 years after its first literary mention in the sixth century CE *Life of St. Columba*.

Setting aside this unresolved issue for the moment, let's assume for sake of argument that we were able to confidently determine that the Ashen Light is a real, physical phenomenon that exists whether or not there are humans on Earth to see it. What, then, would its most probable explanation be? After historical consideration of everything from lightning in the atmosphere to bizarre metallic "snow" precipitating on its loftiest peaks to celebratory conflagrations set by a population of intelligent inhabitants, one by one these ideas have been knocked down for their implausibility and/or lack of supporting evidence. A few tantalizing possibilities, such as ongoing surface volcanism on Venus, remain just sufficiently tenable that they cannot be firmly taken off the list. But one theory rises to the top: nighttime oxygen "green line" airglow in the Venus ionosphere that on occasion emits sufficient visible light that an Earthbound observer with a reasonable telescope stands a better than even chance of detecting it.

As an explanation, airglow checks a lot of the requisite boxes. If the process that yields this light resembles at all the corresponding process in the terrestrial atmosphere—and we have good reason to suspect that, on the whole, it does—then it should be inconsistently distributed both across the Venus disc and through time on virtually all scales. The ubiquity of observations over geography and historical time fits with a recurrent phenomenon like the airglow. It would be most readily visible from Earth when our planet was closest to Venus, as it is around inferior conjunctions of the latter, although in principle it might be observable at any orbital phase. The physical mechanism that powers the Ashen Light is probably most efficient when the Sun is in a magnetically active state and periodically hurling coronal mass ejections toward the planet; however, even in otherwise quiescent times of the solar cycle, interacting solar wind particle streams can do the trick. The airglow might at times become bright enough to trigger some response from the color-sensing circuitry in the human eye and brain, or perhaps they are just filling information where it is expected but not actually detected. There could even be some nexus to "negative" Ashen Light reports, which often take the form of perceived discolorations of the disc against the blue of the daytime sky.

Where airglow fails as a complete explanation of the Ashen Light mechanism is in that it remains ambiguous, yielding no clear documentary evidence. The work of scientists like Tom Slanger and Candace Gray (Chap. 7) shows that the oxygen green line in the atmosphere of Venus can reach intensities rivaling or even exceeding the same emission in the Earth's atmosphere. But it seems unlikely that the failure of observers over recent decades to produce convincing photos of the Ashen Light is the result of simple bad luck or poor timing. There is still something missing that a jury judging this case would rightly point out as a doubt, whether reasonable in nature or in whose shadow the truth may be hiding. The best argument favoring oxygen airglow as the source of the Ashen Light may well be an appeal to a centuries-old problem-solving principle. The idea, usually attributed to the fourteenth-century English Franciscan friar William of Ockham (also spelled 'Occam'), is that when presented with competing hypotheses about the same prediction, one should select the solution with the fewest assumptions.[1] In the case of the Ashen Light, the prediction is contained in the previous list of eight observational facts: whatever might plausibly explain it has to account for all of them. And of the physical mechanisms considered in this book, the one with the fewest assumptions is the oxygen-airglow hypothesis. It has a well-understood terrestrial analog; it is known to operate at some level in the Venus atmosphere; and certain characteristics of light produced in this way account for some of the observational facts quite well.

Bear in mind that we need not adhere dogmatically to Ockham, exactly because nature isn't bound to simplicity as the basis for every physical law. In trying to figure out anything about how the universe works, scientists over the centuries noticed that among the potentially infinite number of possible explanations of a phenomenon, most are hopelessly complex and don't yield to experimentation. Science values the falsifiability of hypotheses; it shows little tolerance for ideas that are intrinsically resistant to trial by (experimental) fire. If an idea isn't subject to being proven *wrong* by empirical test, then it is regarded as equally impossible to be proven *right* through similar means.

[1]Ockham's philosophical "razor" is popularly paraphrased as "the simplest explanation is most likely the right one."

Complex hypotheses are typically more difficult to test than simple ones and therefore more resistant to falsification. It doesn't mean they're necessarily wrong, but rather showing that they're right is usually far more difficult.

In 1955, Gerard Kuiper argued that:

> the observer should be totally disinterested in the outcome of his observation. In some ways this condition sounds paradoxical and impossible; yet it is essential that the observer, once he has decided to make the observation, divorce himself of all wishes, emotions, and recollections, and become an "instrument" till the work is completed.

But although human retinas are electrochemical detectors of light, humans themselves are not "instruments." For us, reality is a mix of stimuli, perception, and interpretation that draws on context, comparison, memory, and predisposition. Our curiosity and intellect have taken us far as a species, including to the edge of what perception can reliably tell us about the world. Beyond that, wishes, emotions, and recollections seem equally competitive motivations.

When I began researching the topic of this book, I feared (subconscious) confirmation bias simply because somewhere deep down, I wanted the Ashen Light to be real. As I researched and wrote, I felt an increasing sense of agnosticism about the topic, knowing firmly only what I didn't really know at all, which is still a lot. The Ashen Light remains paradoxical, shadowy, and mysterious, neither proven nor disproven, ruled in or ruled out, acceptable or accepted. For me, it is as real as the convictions of the reasonable and trustworthy observers through the centuries who said they saw it. Were the Ashen Light nothing but wishful thinking, it would have disappeared long ago with debunked astronomical ideas like the "canals" of Mars and the undiscovered planet Vulcan.

Soon the history of the Ashen Light of Venus will enter a fifth century, and there is no cause to believe that observers will no longer report it. Perhaps new spacecraft visits, astronomical instruments, or data analysis techniques will provide some kind of insight we currently lack, and astronomers of the future will look upon the Ashen Light as a quaint but fully understood part of their collective folklore. Until then, the Ashen Light remains perhaps the most notorious example in the history of astronomy of "the one that got away."

Bibliography

1. Arago, D. F. J., & Barral, F.-A. (1855). *Astronomie populaire* (Vol. 2). Paris: Gide.
2. Barker, R. (1954). Report of the ordinary general meeting of the association held on Wednesday, 25 November 1953, at Burlington House, Piccadilly, W1. *Journal of the British Astronomical Association, 64*(60).
3. Barth, C. A., Pearce, J. B., Kelly, K. K., Wallace, L., & Fastie, W. G. (1967). Ultraviolet emissions observed near Venus from Mariner V. *Science, 158*(Dec.), 1675–1678.
4. Baum, R. (2006). The visibility of the dark side of Venus, 1921–1953: A series of observations by M.B.B. Heath. *Journal of the British Astronomical Association, 116*(Aug.), 190.
5. Baum, R. (2007). *The haunted observatory: curiosities from the astronomer's cabinet.* Buffalo: Prometheus Books.
6. Baum, R., & Sheehan, W. (1997). *In search of planet Vulcan. The ghost in Newton's clockwork universe.* New York: Plenum Trade.
7. Bianucci, P. (1980). Giovanni Virginio Schiaparelli. *L'Astronomia, 6*, 45–48.
8. Black, J. A., Cunningham, G., Fluckiger-Hawker, E., Robson, E., & Zólyomi, G. (1998). *The electronic text corpus of sumerian literature*
9. Blackwell, H. R. (1946). Contrast thresholds of the human eye. *Journal of the Optical Society of America, 36*(11), 624–643.
10. Blackwell, H. R. (1952). Studies of psychophysical methods for measuring visual thresholds. *Journal of the Optical Society of America, 42*(9), 606.
11. Borgato, M. T. (ed). (2002). *Giambattista Riccioli e il merito scientifico dei gesuiti nell'etá barocca: Papers from the conference on the occasion of the 4th centenary of Riccioli's birth held at the Università degli Studi di Ferrara, Ferrara, and in Bondeno, October 15–16, 1998.* Florence: Biblioteca di Nuncius: Studi e Testi.

J. C. Barentine, *Mystery of the Ashen Light of Venus*, Astronomers' Universe,
https://doi.org/10.1007/978-3-030-72715-4

12. Broadfoot, A. L., Kumar, S., Belton, M. J. S., & McElroy, M. B. (1974). Ultraviolet observations of Venus from Mariner 10: preliminary results. *Science, 183*(Mar.), 1315–1318.
13. Brown, D. S. (1970). The ashen light-a refraction effect? *Journal of the British Astronomical Association, 80*(Oct.), 462–464.
14. Clerke, A. M. (1893a). *A popular history of astronomy during the nineteenth century*. London: A. & C. Black.
15. Clerke, E. M. (1893b). *The planet venus*. London: Whiterby and Company.
16. Cox, A. N. (ed). (2002). *Allen's astrophysical quantities* (4th edn.) New York: Springer.
17. Crisp, D., Allen, D. A., Grinspoon, D. H., & Pollack, J. B. (1991). The dark side of Venus - near-infrared images and spectra from the Anglo-Australian observatory. *Science, 253*(Sept.), 1263–1266.
18. Cruikshank, D. P. (1992). The ashen light of Venus. Page 43 of: Edberg, S. J. (Ed.), *Research amateur astronomy*. Astronomical Society of the Pacific Conference Series (Vol. 33).
19. de Heen, P. (1872). De la lumière secondaire de Vénus. *Bulletin hebdomadaire de l'Association Scientifique de France, 11*, 278.
20. Derham, W. (1731). *Astro-theology: or a demonstration of the being and attributes of god, from a survey of the heavens* (6th edn.) London: William Innys.
21. Dobbins, T. (2012). Enigma of the ashen light. *Sky & Telescope*, 50–54.
22. Editors (1876). Our astronomical column. *Nature, 14*(June), 91–92; 131–132.
23. Flammarion, C. (1884). *Popular astronomy*. New York: D. Appleton & Company.
24. Goody, R., & McCord, T. (1968). Continued search for the Venus airglow. *Planetary and Space Science, 16*(Mar.), 343.
25. Gruithuisen, F. P. (1824). *Entdeckung vieler deutlichen Spuren der Mondbewohner, besonders eines collossalen Kunstgebäudes derselben*. Technical Report, Nuremberg.
26. Gruithuisen, F. P. (1842). Astronomische Neuigkeiten aus Beobachtungen. *Astronomisches Jahrbuch für physische und naturhistorische Himmelsforscher und Geologen*, 158–159.
27. Gurnett, D. A., Kurth, W. S., Roux, A., Gendrin, R., Kennel, C. F., & Bolton, S. J. (1991). Lightning and plasma wave observations from the Galileo flyby of Venus. *Science, 253*(Sept.), 1522–1525.
28. Harding, K. L. (1806). Beobachtungen der Nachtseite der Venushügel vom Hrn. Prof. Harding zu Göttingen. *Berliner Astronomisches Jahrbuch*, 167–171.
29. Herschel, W. (1793). Observations on the planet Venus. *Philosophical Transactions of the Royal Society of London, 83*, 209–213.
30. Hersé, M. (1988). Bright nights; past, present, and future trends. Pages 41–64 of: Schröder, W. (Ed.), *Geophysical research*. Interdivisional commission on history of IAGA. Potsdam, Germany: FRG.
31. Houtgast, J. 1955. Indication of a magnetic field of the planet Venus. *Nature, 175*(Apr.), 678–679.
32. Hunt, G. E., & Moore, P. (1982). *The planet Venus*. London: Faber and Faber.

33. Klein, H. (1880). Die Phosphorescenz der Nachtseite der Venus. *Anleitung zur Durchmusterung des Himmels*, 101.
34. Knobel, E. B. (1911). Obituary notices : associates :- Schiaparelli, Giovanni Virginio. *Monthly Notices of the Royal Astronomical Society, 71*(Feb.), 282.
35. Knoll, H. A., Tousey, R., & Hulburt, E. O. (1946). Visual thresholds of steady point sources of light in fields of brightness from dark to daylight. *Journal of the Optical Society of America, 36* (8), 480.
36. Knott, G. (1862). On the stars R Vulpeculae and U Geminorum, and on an appearance in Venus. *Monthly Notices of the Royal Astronomical Society, 22*(5), 157–58.
37. Krasnopolskii, V. A. (1980). Lightning on Venus according to information obtained by the satellites Venera 9 and 10. *Cosmic Research, 18*(Nov.), 325–330.
38. Krasnopolskii, V. A. (1983). *Venus spectroscopy in the 3000–8000 Å region by Veneras 9 and 10* (pp. 459–483). Tuscon: Univ. of Ariz.
39. Krasnopolskii, V. A., Krysko, A. A., Rogachev, V. N., & Parshev, V. A. (1976). Spectroscopy of the nightglow of Venus from the Venera 9 and 10 probes. *Kosmicheskie Issledovaniia, 14*(Sept.), 789–795.
40. Lawrence, G. M., Barth, C. A., & Argabright, V. (1977). Excitation of the Venus night airglow. *Science, 195*(Feb.), 573.
41. Levine, J. S. (1968). On the occurrence of the Ashen Light on Venus. *Planetary and Space Science, 16* (Nov.), 1417–1418.
42. Lorenz, R. D. (2018). Lightning detection on Venus: a critical review. *Progress in Earth and Planetary Science, 5*(1), 34.
43. Lowell, P. L. (1895). *Mars*. Boston: Houghton, Miffilin and Co.
44. Lowell, P. L. (1909). The planet Venus. *Popular Science Monthly, 75* (December).
45. Mazzucato, M. T. (2006). Giovanni Virginio Schiaparelli. *Journal of the Royal Astronomical Society of Canada, 100*(June), 114.
46. Montañés-Rodríguez, P., Pallé, E., & Goode, P. R. (2007). Measurements of the surface brightness of the earthshine with applications to calibrate lunar flashes. *The Astronomical Journal, 134*(3), 1145–1149.
47. Moore, P. A. (1961). *The planet Venus* (3rd edn.) London: Faber & Faber.
48. Muirden, J. (1987). *The amateur astronomer's handbook* (3rd edn.) New York: Harper & Row.
49. Napier, W. M. (1971). The ashen light on Venus. *Planetary and Space Science, 19*, (Sept.), 1049–1051.
50. Newcomb, S. (1902). *Astronomy for everybody*. New York: McClure, Phillips & Co.
51. Newkirk, Jr., G. (1959). The airglow of Venus. *Planetary and Space Science, 1*(Jan.), 32–36.
52. Palczewska, G., Vinberg, F., Stremplewski, P., Bircher, M. P., Salom, D., Komar, K., et al. (2014). Human infrared vision is triggered by two-photon chromophore isomerization. *Proceedings of the National Academy of Sciences, 111*(50), E5445–E5454.

53. Pallé, E., Rodriguez, P. M., Goode, P. R., Qiu, J., Yurchyshyn, V., Hickey, J., et al. (2004). The Earthshine Project: update on photometric and spectroscopic measurements. *Advances in Space Research, 34*(2), 288–292.
54. Palmieri, P. (2001). Galileo and the discovery of the phases of Venus. *Journal for the History of Astronomy, 32*(May), 109–129.
55. Pastorff, J. W. (1825). Fernere Bestätigung, dass Venus, Jupiter und Saturn mit auffallend sichtbaren Lichtsphären umgeben sind, vom Hrn. Geheimenrath Pastoroff, auf Buchholz bei Draußen in der Neumark, unterm 6. Sept. 1822 eigesandt. *Berliner Astronomisches Jahrbuch*, 235–241.
56. Phillips, J. L., & Russell, C. T. (1992). The Venus ashen light – results of the 1988 observing campaign. *Advances in Space Research, 12*(Sept.), 51–56.
57. Rheinauer, J. (1859). *Die Erleuchtung des Planeten Venus durch die Erde*. Technical Report, Freiburg.
58. Rheinauer, J. (1862). *Grundzüge der Photometrie*. Halle: H. W. Schmidt.
59. Riccioli, G. B. (1651). *Almagestum novum astronomiam veterem novamque complectens observationibus aliorum*. Bologna: Self-published.
60. Russell, C. T., & Scarf, F. L. (1990). Evidence for lightning on Venus. *Advances in Space Research, 10*, 125–136.
61. Safarik, A. (1874). On the visibility of the dark side of Venus. *Report of the British Association for the Advancement of Science*, 404.
62. Schaefer, B. E. (1991). Glare and celestial visibility. *Publications of the Astronomical Society of the Pacific, 103*(Jul), 645.
63. Schiaparelli, G. V. (1893). La Vita Sul Pianeta Marte. *Natura ed Arte, 2*(5), 393–404.
64. Schorr, F. (1875). *Der Venusmond: Und Die Untersuchungen Uber Die Fruheren Beobachtungen Dieses Mondes*. Braunschweig: Friedrich Vieweg and Son.
65. Schröter, J. H. (1806). Beobachtung der Nachtseite der Venuskugel, vom Hrn. Justizrath und Oberamtmann Doct. Schröter zu Lilienthal. *Berliner Astronomisches Jahrbuch*, 164–167.
66. Sheehan, W. (1988). *Planets and perception: telescopic views and interpretations, 1609–1909*. Tucson: University of Arizona Press.
67. Sheehan, W. (1997). Giovanni Schiaparelli: visions of a colour blind astronomer. *Journal of the British Astronomical Association, 107*(Feb.), 11–15.
68. Sheehan, W., Brasch, K., Cruikshank, D., & Baum, R. (2014). The ashen light of Venus: the oldest unsolved solar system mystery. *Journal of the British Astronomical Association, 124*(Aug.), 209–215.
69. Sheehan, W., & Misch, T. (2016). A special centennial: Mercury, Vulcan, and an early triumph for General Relativity. *The Antiquarian Astronomer, 10*(June), 2–12.
70. Slanger, T. G., Huestis, D. L., Cosby, P. C., Chanover, N. J., & Bida, T.A. (2006). The Venus nightglow: ground-based observations and chemical mechanisms. *Icarus, 182*(1), 1–9.
71. Smith, E. J., Davis, Jr., L., Coleman, Jr., P. J., & Sonett, C. P. (1963). Magnetic field. *Science, 139*(Mar.), 909–910.

72. Vogel, H. (1873). *Beobachtungen angestellt auf der Sternwarte des Kammerherrn von Bülow zu Bothkamp* (Vol. II). Leipzig: W. Engelmann.
73. Webb, T. W. (1873). *Celestial objects for common telescopes.* London: Longmans, Green & Co.
74. Zenger, C. V. (1883). On the visibility of the dark side of Venus. *Monthly Notices of the Royal Astronomical Society, 43*(Apr.), 331.
75. Zhang, T. L., Lu, Q. M., Baumjohann, W., Russell, C. T., Fedorov, A., Barabash, S., et al. (2012). Magnetic reconnection in the near venusian magnetotail. *Science, 336*(May), 567.

Index

J. C. Barentine, *Mystery of the Ashen Light of Venus*, Astronomers' Universe,
https://doi.org/10.1007/978-3-030-72715-4

B

C

D

K

L

M

S

T

Printed in the United States
by Baker & Taylor Publisher Services